SpringerBriefs in Psychology

Psychology and Cultural Developmental Science

Series editors

Giuseppina Marsico, University of Salerno, Salerno, Italy;
Centre for Cultural Psychology, Aalborg University, Aalborg
Denmark

Jaan Valsiner, Centre for Cultural Psychology, Aalborg University,
Aalborg, Denmark

SpringerBriefs in Psychology and Cultural Developmental Science is an extension and topical completion to *IPBS: Integrative Psychological and Behavioral Science Journal* (Springer, chief editor: Jaan Vasiner) expanding some relevant topics in the form of single (or multiple) authored book. The Series will have a clearly defined international and interdisciplinary focus hosting works on the interconnection between Cultural Psychology and other Developmental Sciences (biology, sociology, anthropology, etc). The Series aims at integrating knowledge from many fields in a synthesis of general science of Cultural Psychology as a new science of the human being.

The Series will include books that offer a perspective on the current state of developmental science, addressing contemporary enactments and reflecting on theoretical and empirical directions and providing, also, constructive insights into future pathways.

Featuring compact volumes of 100 to 115 pages, each Brief in the Series is meant to provide a clear, visible, and multi-sided recognition of the theoretical efforts of scholars around the world.

Both solicited and unsolicited proposals are considered for publication in this series. All proposals will be subject to peer review by external referees.

More information about this series at http://www.springer.com/series/15388

Emiliana Mangone

Social and Cultural Dynamics

Revisiting the Work of Pitirim A. Sorokin

 Springer

Emiliana Mangone
Department of Human, Philosophic and
Education Sciences (DISUFF)
University of Salerno
Salerno, Italy

ISSN 2192-8363   ISSN 2192-8371   (electronic)
SpringerBriefs in Psychology
Psychology and Cultural Developmental Science
ISBN 978-3-319-68308-9   ISBN 978-3-319-68309-6   (eBook)
https://doi.org/10.1007/978-3-319-68309-6

Library of Congress Control Number: 2017954321

Printed on acid-free paper

This Springer imprint is published by Springer Nature
The registered company is Springer International Publishing AG
The registered company address is: Gewerbestrasse 11, 6330 Cham, Switzerland

Series Editors Preface

Interdisciplinarity in the Social Sciences Under Observation: Dissecting the Thought of Pitirim A. Sorokin

There are many good reasons to be proud of hosting Emiliana Mangone's book, *Social and Cultural Dynamics: Revisiting the Work of Pitirim A. Sorokin* in the SpringerBriefs series, *SpringerBriefs in Psychology and Cultural Developmental Science*. We are here to celebrate the link between psychology and developmentally oriented sociology—and Sorokin's contributions were major *tour de force* in twentieth century sociology.

This is, as far as we know, the only one international book to celebrate the 50th anniversary of Sorokin's death, one of the forgotten giants in the history of sociology. And yet, the book is not a mere tribute to his vast production. Instead, it offers a critical and actualized look at Sorokin's theories, with which only very few scholars are familiar.

The intellectual depth of this book surpasses the rigid disciplinary borders in our contemporary academic world. Mangone goes beyond these superimposed limitations and navigates with expertise the waters of other disciplines, first and foremost the always turbulent psychological ocean. Indeed, this book is an interdisciplinary dialogue that follows other examples of the same kind in our SpringerBrief series (Rayner 2017). Both biology and sociology are constructive interlocutors for psychology in its efforts to break out of the behaviourist and cognitivist straightjacket of the twentieth century and to establish its own domain of expertise. Sorokin's ideas—informed by his long life experiences in Russia and North America—make his intellectual efforts worthy of a careful analysis in our time.

Mangone makes an important contribution to a new look at Sorokin's ideas. She carefully documents how Sorokin went about creating sociological principles for understanding the changes of sociocultural systems. Since the declared aims of our SpringerBrief Series is to look at the developmental processes from different theoretical angles, Sorokin's unceasing quest for explaining the basic dimension of the sociocultural dynamics kept our "intellectual sympathy" and thanks to Mangone's accurate work, we also found numerous points of convergence with Cultural Psychology and Sociology. According to Sorokin:

> *"We do not know any empirical sociocultural system or phenomenon which does not change in the course of its existence or in the course of time.* In the whole empirical sociocultural world there has existed hardly any system which has remained unchanged. This observation is incontestable. The objection possible is that though change is unquestionable, it remains unknown to what it is due: to purely immanent forces of the system, or to an incessant influencing of it by a set of external factors" (Sorokin 1998: 241, added emphasis).

So many "unquestionables" and "incontestables"—and yet sociology has had similar trouble in moving to a consistently developmental perspective in the half century since Sorokin's death. Sorokin proposed a third possible answer: an *Integral* perspective that conceives the change of any sociocultural phenomena as the result of the combined external and internal forces. Yet the precise ways in which such combination functions remained unrevealed by him—and continues to be so until our time. Social systems are still viewed as they are (ontologically), rather than in their transformations. There are many descriptions of societal transformations, but no sociological or cultural psychological theory of revolutions has emerged (Wagoner et al. 2017).

The problem raised by Sorokin—that social phenomena exist as relations with environment—was not new in the history of psychology. William Stern (1935), for instance, in his personological account of the human mind sets the *present* boundary between the person and the world as the locus of synthesis of any kind of novelty. Nevertheless, what has been missing here as well as in most of the formal thought models in psychology (and in the social sciences at large) is the time perspective. This is a topic that may benefit from a new look at Sorokin's efforts.

Sorokin made an attempt to introduce an historical and dynamic view of society. His monumental classificatory work about the ideological tendency of temporalism (focus on emergence and development) and eternalism in European society (Sorokin 1985) reminds us of the history of macro-scale societal changes over the centuries. Yet, the micro-scale analysis of the making of such changes remained beyond his scope.

In this vein, as Mangone shows in her interdisciplinary effort, Sorokin's heritage has numerous resonances with what has been developed in the field of Cultural Psychology over the last decades and especially the shift from the epistemology of being to the *epistemology of becoming* in psychology (Marsico 2015; Marsico and Valsiner 2017; Valsiner et al. 2016). This focus on becoming is a central axiom that can open the door for reunification of the issues of memory and imagination as partners in the human progress. In addition, the claim of Sorokin for sociology as a generalizing science of sociocultural phenomena (see Chap. 2) reverberates in Cultural Psychology as the science of human being which develops through careful investigation of the phenomena under study, together with the advancement of high-level abstract generalizations.

Scientific knowledge, in fact, entails generalization that happens on the border of what is already known and what is not yet known.

As Prigogine pointed out:

Our time is one of expectation, of anxiety, of bifurcation. Far from an "end" of science, I feel our period will see the birth of a new vision, one of new science whose cornerstone encloses the arrow of time; a science that makes of us and of our creativity the expression of a fundamental trend in the universe (Prigogine 1996: 42).

Science is a passionate affair. Sorokin's professional and personal life shows how the researcher and his "qualified guessing" is the primary source for any advancement in the knowledge construction. The contrast between being a political prisoner in newly established Bolshevik Russia waiting for his death sentence to be carried out (only to be converted to what was seen as even greater punishment—expulsion from his homeland) and his final years of local fights with fellow Harvard faculty—inmates of a different social enclave—was huge. Sorokin survived both—a tough man as he was. We wish the reader similar stamina to consider the sociological quest that Mangone sets up in her study of Sorokin's legacy. We hope the reading of this Brief will be setting new horizons in the current scientific debates of science, society, and human psychology.

Aalborg, Denmark   Giuseppina Marsico
August 2017   Jaan Valsiner

References

Marsico, G. (2015). Striving for the new: Cultural psychology as a developmental science. *Culture & Psychology, 21*(4), 445–454. https://doi.org/10.1177/1354067X15623020.

Marsico, G., & Valsiner, J. (2017). *Beyond the mind: Cultural dynamics of the psyche.* Charlotte, NC: Information Age Publishing.

Prigogine, I. (1996). Science, reason, and passion. *Leonardo, 29*(1), 39–42.

Rayner, A. (2017). *The origin of life patterns—In the natural inclusion of space in flux.* New York: Springer.

Sorokin, P. (1985). *Social and cultural dynamics* (2nd ed.). New Brunswick, NJ: Transaction.

Sorokin, P. A. (1998). *On the practice of sociology.* In B. V. Johnston (Ed.). Chicago: University of Chicago Press.

Stern, W. (1935). *Allgemeine Psychologie auf personalistischer Grundlage.* Haag: Martinus Nijhoff.

Valsiner, J., Marsico, G., Chaudhary, N., Sato, T., and Dazzani, V. (Eds.) (2016). *Psychology as a science of human being: the Yokohama manifesto.* Vol. 13 in Annals of Theoretical Psychology. New York: Springer.

Wagoner, B., Moghaddam, F., and Valsiner, J. (Eds.) (2017). *From rage to revolutions.* Cambridge, UK: Cambridge University Press.

Acknowledgements

This book had a long preparatory study, yet took a relatively short time to be written. Like all the products of human intellect, it is not the product of the writer's endeavour only. It is my duty to thank those who variously contributed to the realization of this book. Expressing gratitude is always very difficult and making it publicly is doubly so, because I am afraid of forgetting someone and having no other opportunity to amend the oversight. Thanks to my father and brother for their moral and material support and special thanks to my mother for bequeathing me three basic principles of life: respect for humanity, trust and sharing (pooling together) ideas. In writing the latter acknowledgement, my only regret is that, due to her death before the publication of the book, she will never enjoy it.

Special thanks also to Jaan Valsiner, who accepted my request as visiting professor at the Niels Bohr Center for Cultural Psychology, the Department of Communication and Psychology, Aalborg University (Denmark). During my two months in Denmark, this book saw the light also thanks to the fruitful discussion during one of the "Kitchen Seminars" whose topic was one of the central chapters of this work. Thanks to my friend and colleague Pina Marsico who stimulated me—and at the same time "challenged" and "provoked" me—to write this book, and with whom I had a continuous discussion on many of its contents. But I also thank Pina for another reason. We share the same idea: great studies or writings are useless if you do not recognize the Other, because it means to be blind and not recognize the surrounding world and all its transformations. Pina has always recognized, and still recognizes, the other with great and profound respect.

Finally, I would like to thank all the staff members of the University Archives & Special Collections, P.A. Sorokin fonds, from the University of Saskatchewan (Canada) who always answered my requests for documents and clarifications on the references.

Contents

1   **Prologue: The Reasons for a Choice** 1
    1   The Reasons for a Choice 1
    2   Sorokin's Russian Period 3
    3   Sorokin's American Period 4
    4   A Journey with Sorokin 5
    References 8

2   **The Boundaries of Sociology and Its Relation to Other Sciences** 9
    1   Sociology and Its Development as a Science 9
        1.1   The Evolutionary Stages of Sociology 10
        1.2   The Debate on Sociological Knowledge 13
    2   Object and Boundaries of Sociology 15
    3   The Interplay Between Sociology and Other Sciences 18
        3.1   The Interplay Between Sociology and Psychology
            in the American Debate of the Last Century 18
        3.2   Modes of Interplay Between Sociology and Psychology 21
    4   Towards a New Scientific Awareness: Transdisciplinarity 23
    References 25

3   **Integral Theory of Knowledge** 29
    1   The Dyad Knowledge/Reality 29
    2   Sorokin's Sociology of Knowledge 32
    3   The Integralism or Integral Theory of Sociology 35
        3.1   The Systems of Truth 37
    References 39

4   **The Society and Its Paradoxes** 41
    1   Social Universe and Interactions 41
    2   The Construction of Reality and the Interactions (Social Relations) 45
    3   The Paradoxes of Contemporary Society 49
    References 51

5   The Cultural System and the Social Problems. 53
    1   The Cultural Universe. 53
    2   Mentality of Culture (or Culture Mentality). 56
    3   The Superorganic (or Cultural) Phenomenon as a Social Problem 59
    References. 61

6   Personality and Human Conduct 63
    1   The Personality as *Weltanschauung* 63
        1.1   Identity in Contemporary Society 65
    2   Human Conduct and Uncertainty 67
    3   Human Conduct, Trust, and Risk 68
    References. 71

7   From Creative Altruistic Love to the Ethics of Responsibility 73
    1   The Creative Altruistic Love in Sorokin. 73
    2   Beyond Sorokin … the Ethic of Responsibility 76
    3   The Role of Social Sciences and Researchers 78
    References. 80

8   Epilogue: Towards Integral Social Sciences 83
    References. 86

Index. 87

About the Author

Emiliana Mangone is Associate Professor of Sociology of Culture and Communication at the Department of Human, Philosophic and Education Sciences, University of Salerno (Italy). Since 2010, she is associated with IRPPS-CNR of Rome (Institute for Research on Population and Social Policies) and she is a Director of International Centre for Studies and Research "Mediterranean Knowledge." Her main investigative interests are in the field of cultural and institutional systems, with particular attention to the social representations of relationships and knowledge as key elements to the human act, and in migration studies. She recently published the following publications: (with Emanuela Pece), Communication et incommunication en Europe: l'exemple de la représentation des migrants, *Hermès,* 77, 2017; (with Mohieddine Hadhri), Intercultural Complexity of the Southern Mediterranean: Arab-Mediterranean Perceptions and Outlooks, *Journal of Mediterranean Knowledge,* 1(2), 2016.

Chapter 1
Prologue: The Reasons for a Choice

1 The Reasons for a Choice

Deciding to write a book is always a troubled choice for several reasons. Deciding to write this book starting from the theories of Pitirim A. Sorokin can be an academic suicide. Indeed, writing about Sorokin—who was much criticized for being considered a "visionary"—or reflecting on the role of sociology and social sciences in general under the pretext of revisiting the work of this Russian-American sociologist, is to make a difficult choice, especially because, in doing so, deconstruct the establishment and paradigms of this science. Here is the prologue.

The decision to write this book on the 50th anniversary of the death of one of the masters of sociology (in February 2018) has not only a commemorative function. The intent is to highlight how his forgotten theories are actually current indeed, and with deep roots in an unceasing pursuit of an integration of the viewpoints and methods of different human and social disciplines. These roots cannot be uprooted; rather, they should be revitalized by a sociologist, like myself, who since the beginning of her sociological studies believed that science should serve mankind and promote its development.

In these first few lines I will describe the sequence of actions that led me to the choice of writing this book. And now the story begins....

The first time I read about Sorokin was in 1989 and I was studying the book *Masters of Sociological Thought: Ideas in Historical and Social Context* (Coser 1977). What surprised me at the time—now I can understand why—was that, while his name recurred in relation to other authors, the Italian edition (Coser 1977, 1983), unlike the American one, did not include a chapter on him. Even then, this greatly intrigued me: why excluding only a chapter out of the original 15?

As a student, I did not wonder too much about it and went on studying. I was thrilled at having chosen the sociology program, because I thought that this work (both academic and professional) had a public function, as Bourdieu (2013) said much better than I ever could. Reading how these scholars' theories intertwined

© The Author(s) 2018
E. Mangone, *Social and Cultural Dynamics*, SpringerBriefs in Psychology,
https://doi.org/10.1007/978-3-319-68309-6_1

with the history of mankind and with their own biographies (and for Sorokin this was particularly emblematic) fascinated me. My enthusiasm, however, decreased with time, because I found no correspondence between all these theories and a real positive impact on individuals: there was no transition from mere theory to practice. This enthusiasm completely "died" when—after 10 years working as a professional sociologist—I joined the Italian academia. By its very nature, the latter is not trans-disciplinary (Piaget 1972) and, moreover, it does not have a holistic view of society; it definitely does not tend towards the promotion of theoretical innovations, but rather towards the preservation of the so-called traditional approaches. It remains enclosed within the limits of the individual disciplines (for reasons of autonomy, or, more trivially, for problems related to the evaluation and career of researchers) with the result of being self-referential and achieving only a total or partial absence of redefinition of the paradigms, methodologies, and methods. In this way, scientific knowledge does not qualify as an exchange experience resulting from "discussions" and "conflicts" between ontologically different disciplines. And this beyond any real or virtual boundary demarcating the "ranges of motions" of the various disciplines.

I came across Sorokin several times, but every time I was about to write something someone persuaded me not to. I am pointing this out because every choice is situated in time and space. Once I got free from opportunistic intellectuals and scientific constraints, it was time to give voice to my idea of the role and functions of human and social sciences. The latter were lost in chasing the operationalization of the social and human being (see the success of quantitative and test methods), losing sight not only of their peculiar objects of study but also of their aim of serving humanity.

My "silent scream"—as I withheld it for years—found its "voice", its expression, in this book. A voice of denunciation and accusation against a way of "practicing" sociology and other human and social sciences that has essentially forgotten their generative statute. And has forgotten, in particular, that sociocultural phenomena must be studied following their dynamism (in space and time) because their constituent elements (personality, society, and culture) are constantly changing and cannot be studied separately (and this is true not only for sociology). I embraced Sorokin's voice and made it mine, an "inconvenient" sociologist, who since the beginning of the last century denounced the quantophreny of humanities and social sciences in his book *Fads and Foibles in Modern Sociology and Related Sciences* (1956).

But whomever reads Sorokin today must place his thinking within the historical reality that the scholar lived in. He narrates it himself in his autobiography, *A Long journey. The Autobiography of Pitirim A. Sorokin* (Sorokin 1963). The evolution of his thinking saw several phases corresponding to his personal and family events; in a letter to Whit Burnett he says that "Eventfulness has possibly been the most significant feature of my li-adventure. In 68 years I have passed through several cultural atmospheres" (Sorokin 1958a: 178). I will divide the evolution of his thought into two major moments: the Russian experiences, that marked him dramatically, and the American ones, that first made him a renowned sociologist and then sentenced him to oblivion.

2 Sorokin's Russian Period

He was born on January 21, 1889, and lived until the age of 11 among the Komi people, one of the Ugro-Finnish ethnic groups, in the North of Russia. He spent his childhood following his itinerant craftsman father and received his early education, along with a group of other boys, from a peasant teaching people to read and write. When both his father and mother died—and became "independent" but penniless—he began attending an Orthodox college for the preparation of teachers, and ultimately the Petersburg university in a time of great splendour for Russian culture. During this period, he studied law, history, psychology, and sociology, which will become his main discipline. In Russia, he found among his teachers Pavlov, Kovalevsky, Petrajitsky, De Roberty, and Bechtereff, and this resulted in two currents of reference for Sorokin: populist idealism and positivist and determinist behaviourism. Among these masters, those who definitely had greater influence on his thoughts are: Kovalevsky, who in the interpretation of the historical and social phenomena focuses on their multidimensionality, and de Roberty, whom Sorokin includes into the "sociologistic school" because in his sociological and philosophical theories he specifies that the world consists of three major energies: the inorganic one (physiochemical), the organic one, and the social one (superorganic)—a subdivision that Sorokin would resume in his study on social and cultural dynamics.

During his stay in Russia he was imprisoned six times for his political commitment (three times under the Tsarist regime and three times under the Communist regime) and even sentenced to death in 1918. The sentence was not executed on due to Lenin's personal intervention. Sorokin resumed academic activity since the foundation of the faculty of sociology at the University of Petersburg, becoming also its first professor and dean. However, in 1922 he was arrested again for his political activity and exiled by the Soviet government. When, in 1923, Sorokin abandoned Russia and crosses the Atlantic to reach the USA, he was still young, but he had already climbed the social ladder despite many problematic and contrasting experiences (particularly domestic and political ones).

The Russian period was when Sorokin's thought basis was constructed, as it can be seen in his early works in the Russian language as well as in his major works. Due to his wandering life, he learned the basics of reading, writing, and arithmetic without regularly attending school, and with a thirst for knowledge that resulted in reading countless books from classical Russian literature. But the development of his thought can be attributed also to his conversations with what could be considered the *intellighenzia*, particularly priests, deacons, and other famous members of the Orthodox Church. Such teachings led Sorokin to be against both a Philistine condition and a superficial representation of life reduced to mere materiality and sensory perception. To this, however, I must add that Orthodox Christianity also influenced his theoretical approach. In particular, the unitary conception of faith and reason (typical of Russian Christianity) will be found in Sorokin's theory of Integralism: the integration of the systems of truth (meaning, faith, and reason) is the royal road to the knowledge of the entire social reality.

Sorokin himself said that these experiences had been a greater teaching than all the books and conferences. They were attached to a composite existential basis: living on the border between Europe and Asia meant getting acquainted with both social and cultural realities; while for what concerns his personality, the familialism he suffered in his early adolescence clashed and intertwined with his youth experiences of urbanism, social struggle, positivist scientism and mystical knowledge (Sorgi 1975). Despite the complexity of these experiences, when he reached America, as an exile, Sorokin's thought had not yet reached its full maturity.

3 Sorokin's American Period

The American academic circles welcomed him enthusiastically, at least in the early years, so much so that they offered him a chair in Minnesota in 1924 and—after the publication of *Social Mobility* (1927) that made him known to the academic audience—the privilege to found and then direct the sociology department at Harvard in 1930. However, the biggest problem faced by Sorokin just after his arrival in the USA is the approach of American social sciences. In fact, beyond a broad anthropological-cultural horizon and sweeping views of social psychology, the scope of sociology is greatly reduced. It merely performs analysis and research on particular aspects of that society, without worrying about a deeper and more open view on the interpretation of that very social reality through which providing guidance, value, and meaning to the research itself. In other words, the sociology and social sciences that Sorokin found in the USA were characterized by a theoretical poverty that did not correspond to his total conception of man and society. His outlook, sweeping the entirety of humanity and its history, brings him closer to Europeans than to Americans, bringing to fruition his sociological orientations, as well as the psychological, historical, and philosophical ones. He renews both methods and contents through a systematic interpretation of social reality, its structures, and its dynamics (Sorokin 1937a, b, c, 1941a, 1962, 1964). In particular, the four volumes of *Social & Cultural Dynamics*[1] systematically expose Sorokin's conception of society and its social and cultural dynamics. Given the length of the historical period examined and the amount of documentation employed, this study remains unique in the history of sociology. These three works—*Social & Cultural Dynamics*; *Society, Culture, and Personality*; and *Sociocultural Causality, Space, Time*—mark Sorokin's isolation in the American academic world. His criticism of the methods and theories of sociology is nothing more than a critical analysis of the fashion and illusions of American society. With these works, it shifts from the initial enthusiasm following Sorokin's arrival in America, to the annoyance of the power circles of American culture. This gap widens even more when, in his last 20 years, teaching,

[1] We wish to point out that this text has then been published in a version and abridged revised in one volume by the author with the title *Social & Cultural Dynamics. A Study of Change in Major Systems of Art, Truth, Ethics, Law and Social Relationships* (Sorokin 1957).

and research activities, Sorokin focuses fully on implementing a committed sociology. This seems to be a sort of final stage in his life's long pilgrimage and in his intellectual journey: the quest for vital energy that helps humanity overcome its deadly crisis. This energy is found in the power of creative altruistic love. On this theme, Sorokin, in 1949, thanks to funding from Mr. Eli Lilly and the Lilly Endowment can establish *The Harvard Research Center in Creative Altruism*. This centre had the objective to study—in an interdisciplinary way, through the promotion of research and symposia—the theme of altruism, analysing its various types, aspects, and dimensions, as well as the effects on the individual, social, and biological life.

This is the moment in history when Sorokin, from being considered a great scholar on his arrival in America, became, in the eyes of the American culture and academic circles, a prophet, a preacher. At this point the gap between Sorokin and the American culture will never be filled. Here begins the rejection phase: he is ignored by the sociological literature controlled by some groups, his works on structural sociology remain at the foundation of the discipline but are no longer cited. Controversy follows controversy until shortly before his death in February 1968. These polemic debates were not only due to his marginalization by the academic world, but also probably because between him and the American environment there never was a true and proper "fusion". Sorokin always maintained himself on a theoretical plane, disdaining the study of some aspects of American society, which he also heavily criticized in his last works. And America never forgave him.

4   A Journey with Sorokin

What I have achieved with this book is a journey through the Sorokin's works—man and scholar free in thought and action. A sort of intellectual adventure-journey.

Almost all of Sorokin's works represent a more or less long stage in this journey. The longest stages were undoubtedly characterized by *Fads and Foibles in Modern Sociology and Related Sciences* (1956); by *Social & Cultural Dynamics*, in his single volume version (1957), by *Society, Culture, and Personality* (1962), by *Sociocultural Causality, Space, Time* (1964), and by an unpublished and undated document, *The Nature of Sociology and its Relation to other Sciences* (Sorokin n.d.) of which the University Archives & Special Collections, P.A. Sorokin fonds of the University of Saskatchewan (Canada) graciously lent me a copy. Although it has not been possible to accurately date this document, the references in it suggest with some degree of certainty that it is later than 1929 but earlier than the publication of the first volume of *Social & Cultural Dynamics* (Sorokin 1937a), as it does not cite any work by Sorokin published in this period, either alone or with co-authors. This manuscript probably represented the "canovaccio" from which all Sorokin's work after *Social Mobility* (1927) have sprung—expanding and developing the reflections in this document.

The journey with Sorokin's gargantuan work meant not only reading this scholar's basic concepts, but represented also an opportunity to try to update them, making them more akin to contemporary society, and to re-affirm the duty that sociology, as well as and other humanities and social sciences, has towards mankind. Indeed, this journey does not lack insights and incursions into other sciences, such as psychology and, in particular, cultural psychology (Valsiner 2014, 2017) that has many similarities with Sorokin's theoretical orientations. Clearly, the journey has necessarily required the selection of some concepts and issues rather than others. I leave readers to delve deeper into the topics that this books outlines, but for which the necessary references are provided.

The journey is eight stages long, and it begins with a prologue (Chap. 1) in which I clarify the reasons for this choice. The Chap. 2 is devoted to sociology, its limitations, and its relation to other humanities and social sciences. The goal of this stage is to explain, in general terms: (a) the evolutionary phases of this discipline of which Sorokin himself sketches the key elements in *Contemporary Sociological Theories* (1928) and *Sociological Theories of Today* (1966); (b) the debate on autonomy, addressed in *Fads and Foibles in Modern Sociology and Related Sciences* (1956); (c) the definition of the object of study of sociology and its limits (intended as boundaries); (d) the relation to the other natural and social sciences, particularly psychology. The last two aspects are covered in *Fads and Foibles in Modern Sociology and Related Sciences* (1956) as well as in *The Nature of Sociology and its Relation to other Sciences* (n.d.).

The third stage represents the kernel of Sorokin's entire work: Integralism (Chap. 3). Sorokin's Integralism—synthesized in *Integralism is My Philosophy* (Sorokin 1958b)—highlighted how, even today, knowledge can influence the construction of reality, and at the same time the development of society and the relationships between individuals. This section states the need for a reflexive, integral theory of knowledge. Being reflexive means being able to promote the construction of the connections in the living environment of individuals and between individuals. All this must be done acknowledging the autonomy of individual disciplines in the social sciences (sociology, psychology, anthropology, etc.), but at the same time abandoning any self-referential excess that exhausts all possible knowledge within its own frameworks and paradigms.

The three subsequent stages address Sorokin's indivisible sociocultural trinity (1948)—Society, Culture, and Personality. Society (Chap. 4), being the result of the interactions of individuals from which relationships and sociocultural processes arise (superorganic phenomena), must be studied by observing the interaction system implemented by individuals. In this chapter, I took the opportunity to explain some features of contemporary society, including its paradoxes, through the analysis of the forms of interactions as they had been defined and described by Sorokin. Another key element is the culture, formed by both objective and subjective elements (Chap. 5). All activities and institutions are "cultural", since their functioning requires meanings to be made explicit. Culture is therefore a key component for the actions of individuals. In these processes, the two dimensions of time and space are paramount. And it is precisely along the temporal and spatial axis that Sorokin describes integrated cultural systems—in detail and with innumerable examples. Integrated cultural systems arise from the various successive cultural mentalities

(Ideational, Sensate, and Idealistic) that have appeared in the course of history (theory of cyclical movements).

In the final part of this stage, there is my attempt to outline a model—adapting Griswold's "cultural diamond" (1994)—that shows the connections and the relationships developed between the elements involved in the construction/production of social phenomena (superorganic phenomena) as cultural objects in their form as social problems. The last of these three stages (Chap. 6) is the personality which is the subject of interaction (individual or collective). The three components of personality (body, mind, and soul) in Sorokin's theories correspond to the three means for knowledge (empirical-sensity, reason, and intuition) and constitute its essence. This dialectic process establishes a direct relationship between the dominant mentality of society and the subject's conduct within it. This relation, however, does not result in a one-to-one correspondence. It is very variable in some societies and more explicit in others. In the second part of this stage, I will introduce an updated reading of what Sorokin claims and argues about personality. The aim is to provide an overview of how all the elements (personality, society, and culture) involved in different processes interconnect and how they permeate their own changes in a highly uncertain society.

The final stage (Chap. 7), that concluded my journey through Sorokin's works, is what I have wanted to title *From creative altruistic love to the ethics of responsibility*. The end-point of Sorokin's strenuous work is "reduced" to an *intuition* that may strongly be summarized in the following statement: the future of mankind and its development it is in the hands of mankind itself. And in order to determine favourable conditions for the peaceful development of mankind, we need to find some way to strengthen solidarity through the positive actions of individuals (*creative altruistic love*). Neither law nor education, let alone religion, economics, or science, is enough for the task. This difficult task belongs to humanity. According to Sorokin, change must start from the rediscovery of man's positive values, and science arises as a guide also by overcoming of strictly sensuous knowledge models. Sociology, therefore, is not merely a sociology of the crisis (Sorokin 1941b) but rather a "critical sociology". It does not merely analyse society's degeneration, but rather searches for its deep roots, denouncing the negative factors causing it. Applying these principles implies understanding the mechanisms through which human beings make their own decisions. And these dynamics highlight the problem of the choice and of how people choose (ethics), as well as the role of the researcher.

The conclusion of this journey—in many ways an adventure—brings me back to the starting point.

The epilogue (Chap. 8) is the need that sociology and other humanities and social sciences regain the leading role in the development of mankind. They can do so only by implementing a comprehensive system of knowledge, analysis, and research that is also reflexive. In these research activities, the researcher must never forget that she is part of the reality that she is studying and that it cannot be otherwise. She must place herself in a reflexive position.

And now I can only wish good reading to those wanting to venture into this journey.

References

Bourdieu, P. (2013). In praise of sociology: Acceptance speech for the gold medal of the CNRS. *Sociology, 47*(1), 7–14.

Coser, L. A. (1977). *Masters of sociological thought. Ideas in historical and social context.* New York: Harcourt Brace Javanovich.

Coser, L. A. (1983). *I maestri del pensiero sociologico.* Bologna: Il Mulino. (Original work published 1977).

Griswold, W. (1994). *Cultures and societies in a changing world.* Thousand Oaks: Pine Forge.

Piaget, J. (1972). L'épistémologie des relations interdisciplinaires. In OCDE (ed.), *L'interdisciplinarité: problèmes d'enseignement et de recherche dans les universités.* Paris: OCDE. Retrieved June 20, 2015, from http://www.fondationjeanpiaget.ch/fjp/site/textes/VE/jp72_epist_relat_interdis.pdf.

Sorgi, S. (1975). La sociologia integrale di P.A. Sorokin. *Sociologia, 1*, 5–47.

Sorokin, P.A. (n.d.). *The nature of sociology and its relation to other sciences.* University of Saskatchewan: University Archives & Special Collections, P.A. Sorokin fonds, MG449, I, A, 3.

Sorokin, P. A. (1927). *Social mobility.* New York: Harper.

Sorokin, P. A. (1928). *Contemporary sociological theories.* New York: Harper.

Sorokin, P. A. (1937a). *Social & cultural dynamics. Vol. I: Fluctuation of forms of art.* New York: American Book Company.

Sorokin, P. A. (1937b). *Social & cultural dynamics. Vol. II: Fluctuation of systems of truth, ethics, law.* New York: American Book Company.

Sorokin, P. A. (1937c). *Social & cultural dynamics. Vol. III: Fluctuation of systems of social relationships, war and revolution.* New York: American Book Company.

Sorokin, P. A. (1941a). *Social & cultural dynamics. Vol. IV: Basic problems, principles and methods.* New York: Bedminster.

Sorokin, P. A. (1941b). *Crisis of our age.* New York: E.P. Dutton.

Sorokin, P. A. (1948). *The reconstruction of humanity.* Boston: The Bacon.

Sorokin, P. A. (1956). *Fads and foibles in modern sociology and related sciences.* Chicago: Henry Regnery.

Sorokin, P. A. (1957). *Social & cultural dynamics: A study of change in major systems of art, truth, ethics, law and social relationships.* Boston: Porter Sargent.

Sorokin, P. A. (1958a). A philosopher of love at Harvard. In W. Burnett (Ed.), *This is my philosophy. Twenty of the world's outstanding thinkers reveal the deepest meaning they have found in life* (pp. 178–180). London: George Allen & Unwin.

Sorokin, P. A. (1958b). Integralism is my philosophy. In W. Burnett (Ed.), *This is my philosophy. Twenty of the world's outstanding thinkers reveal the deepest meaning they have found in life* (pp. 180–189). London: George Allen & Unwin.

Sorokin, P. A. (1962). *Society, culture, and personality: Their structure and dynamics: A system of general sociology.* New York: Cooper Square.

Sorokin, P. A. (1963). *A long journey: The autobiography of pitirim A. Sorokin.* New Haven: College and University.

Sorokin, P. A. (1964). *Sociocultural causality, space, time: A study of referential principles of sociology and social science.* New York: Russel & Russell.

Sorokin, P. A. (1966). *Sociological theories of today.* New York/London: Harper e Row.

Valsiner, J. (2014). *An invitation to cultural psychology.* London: Sage.

Valsiner, J. (2017). *Mente e culture: La psicologia come scienza dell'uomo.* Rome: Carocci editore.

Chapter 2
The Boundaries of Sociology and Its Relation to Other Sciences

1 Sociology and Its Development as a Science

It is impossible to approach a revisiting of Sorokin's works without a background knowledge of the evolution of his science, i.e. sociology. This chapter provides a general overview of the evolution of sociology, its subject of study, and its boundaries, also in relation to other sciences.

Several voices argued that sociology was a new science that came to the forefront at the end of the nineteenth century. Its aim was to study the changes in the forms of association of people and institutions that had been observed through this century. Those same changes that later characterized the so-called "modern states". But if we want to really explain sociology, we must look more closely at its origins. It is customary to date the birth of sociology when Comte, in his book *Cours de philosophie positive* (1830) employs the term "sociology"[1] for those studies which had so far been termed "social physics". However, sociology has existed for much longer than it is commonly imagined, although not as a science. One need only consider the countless philosophical studies that contained the first "sociological thought". Studies that, over the centuries, have dealt with societal transformations and the relationships between social structures and individuals (for example, Aristotle's study of the birth of Greek cities). After all, it is not conceivable for a science to be born out of nothing: no significant cultural phenomenon originates in a social void. Hence the most basic definition of sociology as: *the science of society*.

It is acknowledged that there is no unique way of temporally dividing the history of sociology, since it is usually related to the formation of the scholar trying to explain it. We thus refer to the extensive literature on the history of sociological thinking (for example: Collins 1988; Coser 1971; Sorokin 1966; Wallace and Wolf 1991). Sorokin (1962), for example, distinguishes the following periods in

---

[1] The new term results from a neologism combining two words: the first of Latin origin (*socius, societas*), the second of Greek origin (*logos*).

© The Author(s) 2018
E. Mangone, *Social and Cultural Dynamics*, SpringerBriefs in Psychology,
https://doi.org/10.1007/978-3-319-68309-6_2

sociological development: "In the Orient, Greece, Rome, Medieval Europe, and Arabia", "From the Renaissance to Modern Times" and "Recent and Contemporary Sociology".

The question, therefore, is: which changes occur, at the end of the nineteenth century, in the studies on social transformations?

What happens is that, on the one hand, the religious aura that had accompanied the reading of the transformations of primitive societies disappears, and, on the other hand, new research methods—developed mainly for natural sciences—are adopted. For these reasons, in this chapter we will not discuss the history of the term and its dissemination, nor will we refer to contemporary sociologies. We will propose some observations on the evolution of sociological thought, the object of study, and the boundaries of this discipline, also in relation to other social sciences, particularly psychology. These are also the fundamental elements on which Sorokin bases all his considerations on sociology as a science.

Nisbet (1977) suggested two ways of presenting the history of sociological thought. The first is based on thinkers whose works are the very content of this discipline; the second is based on the *schools*. In this work, we will try to integrate both methods. It is important not to limit the analysis to the most authoritative figures in a given science, but to take also into considerations its more widespread orientations and traditions, as these contributed to the affirmation of the autonomy of the discipline as a science and to a critical reflection on it.

What characterized the evolution of sociology was not so much its object of study, which had been fairly clear since Comte, but rather the need to make it autonomous from other sciences and especially from natural sciences. However, a clarification about the subject of sociology is essential.

1.1   The Evolutionary Stages of Sociology

Sorokin stated that

> "the social sciences are concerned with *superorganic* phenomena. As the presence of life distinguishes of mind or thought in its development form, differentiates the superorganic phenomena from organic. [...] Superorganic phenomena so developed are found only in man and the man-made world. Other species exhibit only rudimentary forms of the superorganic. Sociology and the other social sciences consider man and the man-made world only with reference to superorganic mind or thought. [...] The task of sociology and social science begins where the physical and biological study of man and his world ends" (1962: 3).

Sorokin, in an undated manuscript,[2] defines

---

[2] *The Nature of Sociology and its Relation to other Sciences,* is an unpublished, undated manuscript by Sorokin, of which the University Archives & Special Collections, P.A. Sorokin fonds, University of Saskatchewan (Canada) kindly lent me a copy. Although it has not been possible to precisely date this document, the references cited in it suggest with some degree of certainty that it is later than 1929 but earlier than the publication of the first volume of *Social & Cultural Dynamics* (Sorokin 1937), as it does not mention any of Sorokin's works published in this period, either alone or with co-authors.

"sociology as the science of the phenomena of human interaction, their factors and results. The term 'interaction' [...] includes the mutual interaction of two or more individuals, as well as the action of one individual upon another. Neither the phenomena of the interaction of inorganic or their constituent parts, nor those of living organism (exclusive of man) are included in the sphere of sociology. Sociology is a science dealing with human interaction only; it is essentially a study of *men* in their interactions—a *homo*-sociology" (Sorokin n.d.: Chap. I, pp. 1–2).

This is the demarcation line highlighting that the

"difference between the tasks of sociology and those of the other generalizing social sciences stands the important difference in their fundamental presuppositions concerning the *nature of man and the interrelations of social phenomena*. [...] In contradistinction of these presuppositions, *homo socius* of sociology is viewed as a generic and manifold *homo*, simultaneously and inseparably economic, political, religious, ethical, and artistic, partly rational and utilitarian, partly nonrational and even irrational, with all these aspects incessantly influencing one another. Consequently each class of sociocultural phenomena is viewed by sociology as connected with all the other classes (with varying degrees of interdependence), as influenced by and affecting the rest of the sociocultural universe. In this sense the sociology studies man and the sociocultural universe as they really are, in all their manifoldness [...]. From these two fundamental differences follow several others distinguish the essential principles and methods of sociology from those of the other social sciences" (Sorokin 1962: 8).

In all his works, Sorokin will continue to claim that this is the subject of study of sociology—an idea that the author of this book fully shares. Sorokin often criticized—quite undiplomatically—his colleagues, both for using the methods pertaining to natural sciences (Sorokin 1956) and for their subject of research.

Sorokin himself had previously synthetically expressed these same statements. In an article, he states that "sociology is and should be a generalizing science". He also elucidated the difference with historical sciences: "*sociology is interested only in those aspects of social phenomena and their relationships which are repeated either in time or in space or in both; which consequently exhibit some uniformity or constancy or typicality*" (Sorokin 1931: 23). Sorokin also differentiates *general sociology* from the *special sociologies* that will be enumerated and specified further with the definition of sociology: "a generalizing science of sociocultural phenomena viewed in their generic forms, types, and manifold interconnections" (Sorokin 1962: 16).

If we are indebted to Comte for coining the term sociology, its dissemination is due to the transformation of the traditional forms of social life triggered, on the one hand, by the French Revolution and, on the other, by the Industrial Revolution.

Towards the twentieth century, the history of sociology reaches an important stage by virtue of Émile Durkheim's works. It was this author's intention to build a social science that could act as a solid foundation for public action, while being well aware that the budding sociological research still had major limits. However, Durkheim's methodological innovation is consequential. He separates *individual* from *society*, laying the foundations for the approach that will eventually become dominant with Parsons' functionalism (1937) and with Merton's version of functional analysis (1968). Society prevails on the individual and gains meaning through institutions that represent stability compared to the variability and mutability of individuals. Society cannot be explained through individual actions and motivations, but through *social facts*.[3] Even phenomena typically considered as individual ones, such as suicide (Durkheim 1897), have a social determiner. At the heart of Durkheim's sociological analysis is the *social fact*, that differs from the psychic fact. Social facts originate not from individual personality, but from a different environment, and influence it in a specific way. Durkheim states that the individual becomes an integral part of the social whole, *the organ of an organism*, only after having bested his egocentric nature. In Durkheim's theorization, sociology must therefore identify and oust possible disagreements within society before they generate discomfort among its members. It should also assume a guiding role for individuals, leading them to plan their behaviours within a society that is increasingly functionally differentiated.

If Durkheim asserts that the structure prevails over the individual, Max Weber, another classical author, considers sociology as the comprehensive science of social action. Weber's (1978) "sociology of meaningful understanding" (*Verstehen*) qualifies human action as a social action only when it is "meaningful". Actions are social when individuals take into account the actions of others, which are driven by *individual motivation*. Therefore, actions are social as they always refer to the behaviour of others, which in turn influences them in their evolution. In other words, social actions must be defined in terms of "objective meanings" of the activity of the social actor. And, in this perspective, sociology "is a science which attempts the interpretative understanding of social action in order thereby to arrive at a casual explanation of its course and effects. In 'action' is included all human behaviour when and in so far as the acting individual attaches a subjective meaning" (Weber 1947: 88). For Weber, actions become the key for the interpretation of Western modern society, which is increasingly dominated by instrumental rationality.[4]

Durkheim and Weber marked the beginning of those that are known in sociology as, respectively, the "French school" and the "German school". Since the 1930s, however, the social sciences in the United States have experienced a strong boost, that will lead to establishing a sort of supremacy of North American thought in sociology—the "americanization" of social sciences (Manicas 1987). Sorokin has certainly contributed to this supremacy of American sociology from his arrival in

---

[3] It should be emphasized that Durkheim (1895) refers to social events as a reality *sui generis* and defines them as ways of acting, thinking, and feeling capable of existing outside of individual consciences. These types of behaviour or thought are external to the individual, coercive, and general.

[4] Weber defines a type of social action (goal-instrumental and value-rational actions, affective action, and traditional action) through the conceptual tool of the *ideal type*. Weber's development of the ideal type begins with the criticism of the use of collectivist concepts. In fact, he thinks that sociology should proceed from the actions of the individual, few or many separate individuals, and the concept of the ideal type is nothing more than a tool that allows to measure reality. It is not an accurate copy of reality; it is only a means that emphasizes the connections that the scholar considers relevant. Through the ideal type of a phenomenon, the researcher will then focus on those connections that have meaning for her in the causal web that determines the phenomenon itself.

the United States (1924) and until he did not meet his colleagues' opposition due to his criticism of American society and of their idea of sociology and sociological methods.

1.2   The Debate on Sociological Knowledge

In light of these basic evolutionary "stages" of sociology, we can understand how sociological knowledge is paramount for a concrete and effective reading of social phenomena. The issue, however, is to be able to build and maintain significant correlations between the sociological thinking and its autonomy from other sciences without downplaying, all the while, the need for integration and disciplinary interdependence. The debate on the autonomy of sociology has been lively since its inception. It branched in two opposing currents of thought: one hoped for a discipline closely linked to natural sciences through the adoption of their empirical methods (positivism); the other argued the total autonomy of sociology, which could not withstand procedural contamination in scientific investigation (interpretive method).

These are still the terms of the eternal quarrels between positivist empirical methods and interpretive ones. If we consider the various historical periods and different cultural environments within this *querelle*, there was no lack of harsh criticisms:

> "Any science, at any moment of its historical existence, contains not only truth but also much that is half-truth, sham-truth, and plain error. This has been especially true of the social and psychological disciplines, for the complexity of mental and social phenomena allows many a fallacy to be taken for the last word of science, 'operationally defined, empirically tested, and precisely measured'. Even the sociology and psychology of today are not exceptions to this rule. They, too, contain verities; they, too, are contaminated by the diseases of sham-truth and error. Some of the ailments are well hidden in the recesses of their valid propositions while others infect their methods, techniques and tests" (Sorokin 1956: v).

It is now ordinary to speak of an uneven development of sciences for what concerns both their internal development and the relations between them, especially after Thomas Kuhn's work *The Structure of Scientific Revolution* (1962). This general idea attributes to social sciences—including, of course, sociology—the bottom rung in the hierarchy of development. These disciplines, it is claimed, lack what Kuhn calls a *paradigm*.[5] For sociology, however, this "deficit" in paradigms is not akin to a lack of theories and methods for scientific research. On the contrary, we are faced with a vast amount of "weaker" paradigms, none of which emerges as hegemonic, a condition that, despite Kuhn's claims, stimulated the development of the discipline. Scholars have even focused on the crisis of sociology, bringing on the

---

[5] The term *paradigm* defines a system of concepts to which is attributed the function of guiding and organizing a scientific research, so as to make it immediately communicable and alterable within the community.

spiralling process that many defined as *the sociology of sociology* (Morin 1984). This situation has in turn created a new genre in sociological production starting from the late 1960s. In short, the need to find a sociological paradigm that prevailed over the others has been felt since the birth of the discipline in the nineteenth century.

The terms of the debate remain these even if we consider different historical periods and contexts, such as the North American one (between ethnomethodologists and theorists of survey analysis) or the German one (Frankfurt School). Such debates, which have sprung all over the world, tend to have such high levels of generality and so poor empirical references that it often becomes difficult to identify what consequences may derive from the various positions in terms of research methods. In recent years, reflection on the theory of the science of society has experienced great stimulus and impetus. The research is partly oriented on well-tested itineraries, while for many others it proposes the variations needed to overcome an unwholesome stagnation. After all, it is not the first time that sociology had to face great discomfort, due essentially to its epistemological limitations, which prevented a reading—at least general—of society (Gouldner 1970). Despite all these difficulties in reading society (or societies) and social phenomena, the need for knowledge is a distinctive element.

The controversy on the autonomy of social sciences still witnesses the opposition of the two currents of thought. However, today one can argue that both approaches have a common element: the conception of sociology as a science emancipated from natural sciences. In his *Declaration of Independence of the Social Sciences* (Sorokin 1941), Sorokin expresses a clear position on sociology and social sciences in general:

"Sociology and social science of the first half of the twentieth century were obsessed by an ambition to be copies of the natural sciences" (Sorokin 1941: 222). A little further on, he continues: "Sociology and the social sciences will abandon their insane ambition to be pseudo-mechanics, pseudo-physics, or pseudo-biology. They will reclaim their lost primogeniture to be a science studying socio-cultural phenomena directly, with their own system of referential principles fitted to the peculiar nature of socio-cultural reality. More specifically, this reformation means the following fundamental changes in sociology and the social sciences of the first part of the twentieth century. A. They will *fundamentally revise their system of truth and knowledge*, [...] This truth of senses must be replaced by a more adequate *integral system of truth, consisting of the organic synthesis of the truth of senses, truth of reason, and truth of intuition, each mutually checking and supplementing the others*. [...] B. The system of truth changed, the framework of the referential principles and methods of the social sciences must change also. [...] C. When the system of truth is changed and the framework of referential principles and methods is transformed, the content and character of the social sciences will be fundamentally modified from top to bottom. [...] D. The importance of such knowledge for applied social science is obvious. [...] When this revolutionary transformation has been made, then, and only then, will the social sciences be real, autonomous scientific disciplines, truly equal to the natural sciences. Then, and only then, will they embody real knowledge and true wisdom. Then, and only then, will they be a perennial value as immortal as any real value. Therefore, this Declaration of Independence of the Social Sciences is also their Magna Charta. It aims to elevate them from the status of serfdom to the natural sciences and to politicians up to a status of equal membership with other sciences and with philosophy, religion, art, and ethics in the great union of the cardinal socio-cultural values" (Sorokin 1941: 226, 228, 229).

For many years still, the terms of the quarrel between empirical-analytical methods (with emphasis on quantity and measure), and hermeneutic-interpretive methods (with emphasis on subjective meanings and qualities) will remain the same. They still look the same despite some scholars' attempts to overcome them—without apparently having achieved the goal. These include: Habermas's theory of communicative action (1984, 1987) based on anthropological conception and rationality; Parsons' voluntary approach (1937), based on the interpenetration of the various systems of his model (AGIL), and that of Alexander (1982), based on a multidimensional vision of social phenomena; and, finally, Giddens' theory of structure (1979), which attempts to combine two usually opposed elements (action and structure). Obviously, qualitative methods and quantitative methods continue to coexist and to complement each other in some studies. They are not necessarily at opposite poles, and adopting one does not mean excluding the other. They offer the opportunity to look from "different angles" at aspects of the same social phenomenon, allowing for a more effective reading of its complexity.

The "debate of sociology about sociology" does not focus anymore on overcoming of this dispute between empirical-analytical methods and hermeneutical-interpretive ones—as their integrated coexistence is taken for granted. Rather, it concerns its object of study, the conjugation between theory and empiricism, as well as its relation to other disciplines. The idea of a supremacy of quantitative methods over qualitative ones has nowadays been overcome, with the acknowledgement that it is possible to know also what cannot be measured with mathematical modules or models.

2 Object and Boundaries of Sociology

In light of the above considerations, sociology and all other social sciences are starting to assume a more prominent position in the general scientific panorama, even though they do not attain the "accuracy" of natural sciences. The cause-effect principle is replaced by that of multicasuality, characterized by the presence of significant relations and dialectical reciprocity between causes and effects (several conditioning factors). It follows that,

> "To define the field of sociology, as with any science, means to select the category of facts that are the object of its study—in other worlds, to establish a special point of view on a series of phenomena that is distinct from the point of view of other sciences. No matter how diverse the definitions by means of which sociologists characterize the existence of social or superorganic phenomena, all of them have something in common, namely, that the social phenomenon—the object of sociology—is first of all considered the interaction of one or more kinds of center, or interaction manifesting specific symptoms. The principle of interaction lies at the base of these definitions; they are all in agreement on this point, and their differences occur further on, regarding the character and form of this interaction" (Sorokin 1998: 59).[6]

---

[6] This essay had been previously published as a pamphlet in Russian in 1913 by Obrazovanie in St. Pertersburg.

Since society is becoming differentiated and increasingly complex, we are faced with new forms and dimensions of solidarity (Zoll 2000). Everyone's life depends on the products of others, and this mutual dependence leads to organic solidarity. With the passage—typical of modernization—from mechanical solidarity to organic solidarity (Durkheim 1893), the "solidarity function" moves from "social norm" to "social work" (aid as a legitimate and paid social function in which the object of exchange is aid itself). If we consider social research activities in general, we can state that they study the ability to act and the ways in which that part of subjectivity or intersubjectivity interacts with the social system. In other words, the work of the sociologist must combine system (objective dimension) and individuals (subjective dimension), that is, it must be able to combine objective and subjective aspects. Sorokin himself, despite maintaining a very critical position on methodologies, assumes a "conciliatory" position in the integrated use of "objectivism" and "subjectivism":

> "We are deeply convinced, that sociology can attain great results in the study of social phenomena through this method, but our recognition of its importance does not indicate that we advocate the complete abandonment of the methods of subjective psychology. On the contrary, we do not regard as utterly futile the use of subjective psychic states in the interpretation and comprehension of social facts. Ours is a more eclectic point of view: we believe that in the study of the phenomena of interaction all the methods which are helpful in any way for the understanding of a given phenomenon should be used without hesitation" (Sorokin n.d.: Chap. III, pp. 14–15).

In contemporary society, all this must be combined with the crisis of the systems and the attempts to define and introduce new policies. These attempts have not forestalled the fragmentation of legal protections, nor the deterioration of the social fabric that needs to be reconstructed through the realization of new forms of solidarity for the citizens' well-being. And it is precisely in this reconstruction process that the work of the sociologist is positioned (which can be understood as an intermediate position between the civil role and the political role). Sociologists should pay attention to all these aspects of change and not just to those pertaining to specific areas of society, and they should also reflect on their own activities.

> "The conclusion [...] is that the basic method for the study of interaction is the objective, which deals only with the external and observable facts of functional interdependence in the behavior of men. In addition, for the sake of a fuller understanding of social phenomena, in many cases introspection may be employed as a subsidiary method. Sociology has to be objective—a science of observable facts and processes—but the results obtained by the objective method, to be more fully exact, should be supplemented by data obtained by the methods of subjective psychology" (Sorokin n.d.: Chap. III, p. 19).

Sociological knowledge helps breaching the wall represented by the complexity of problems and situations, and allows for a better conjugation of objective and subjective dimension. This engenders a series of problems that can only be overcome if in sociological works the "knowledge is transferred and not ignored".

Sociologists are heavily involved in this dual value, which they find hard to "break free" from: on the one hand, they are "institutional accompanists"; on the other hand, they are "critical and active citizens", analysts and subjects of analysis at the same time. Sociologists do not seek to aseptically understand the problems;

rather, as part of the society, they believe themselves involved. They do not seek to defend themselves from society, preferring to build a relationship between the actors to make society more "tailored" for all citizens. Knowledge, which must be associated with action, does not seek the solution but the possible paths to be taken to better the problematical issue. Sociological work is thus political, not in the sense of party membership, but of having political weight as social actors, and therefore bearers of values, meanings, and subjective and social rights. Through their continuous research for knowledge, inequalities are recognized.

Consequently, the object of study of sociology is the individual and collective phenomenological reality in relation to social systems. Sociological study cannot therefore be oriented only to *macro-social* phenomena (relating to social systems and forms of organization), to the exclusion of *micro-social* phenomena (concerning individual/society relationships and social actions) or *meso-social* ones (concerning the relations between the social system and the lifeworld, where the latter is understood as the set of meanings and representations of the culture). Following this perspective, we can state that the "world sociology" that strongly developed with the globalization processes will have to resume criticism of social representations. Indeed, sociology seems to have difficulties in reading social changes due to its excessive self-referentiality or, in other words, its insularity with respect to other disciplines, which makes it exhaust all possible knowledge within its own reference frameworks and paradigms. This isolation from other disciplines (insularity) is a criticism that Sorokin always aimed at to sociology. To this, we can add another problematic issue, namely that sociology has lagged behind in reaching the standards that other disciplines (e.g. psychology and economics) have achieved in integrating theory and empirical research. According to Goldthorpe (2000), this condition (defined as the "scandal of sociology") is aggravated by the fact that sociologists have not even realized the direness of this situation. Goldthorpe's criticism is briefly expressed in these terms: contemporary sociologists are clearly divided on the relationship between their two main activities, empirical research and theory. In addition, they are also divided on what kind of academic or scientific enterprise is or should be sociology. Finally, there are significant differences in how to interpret and respond to this situation of intellectual division or, better, of disciplinary fragmentation. Obviously, it is not simple to respond to these criticisms, after all it is the goal of this work. However, considering a logic where social research activities are placed in an integrated perspective, there is no boundary between scientific research, professional activity, and social usability.

In light of these considerations, we can no longer speak of a contrast between theory and empiricism, or between the objective dimension and the subjective one. We are faced with a continuum of interdependencies ranging from theory to empiricism, up to operativeness, and between the objective (and objectivable) aspects of social reality and the subjective ones of symbolic mediation. Empirical research thus becomes essential for acquiring knowledge, which in turn should allow for a reading of social (individual and collective) phenomena, in order to translate theoretical premises into actions that are not merely technical, but also reflexive on the activities themselves. Sociological knowledge is therefore the instrument that allows for a holistic understanding of the complex network of problems related to social phenomenology. In other words, sociological knowledge should be gained through systematic and methodologically founded observation, deemed the main activity to overcome Comte's "social physics", and to lay the foundations for an intervention able to entail modifications/transformations both at the individual and at the social level. It is therefore necessary to try to redefine the paradigms of sociology in a direction that takes in account of both its different dimensions and the other social sciences. A sociology that does not integrate the contexts within which the actions take place and the individual or individuals acting is not conceivable, nor is it possible. The search for the reasons why phenomena occur should no longer be limited only to the *cause*, but to a *meaning*.

3   The Interplay Between Sociology and Other Sciences

The relationship between sociology and the other sciences, especially physics and natural sciences, clearly depends not only on the object of study, but also on the methods of analysis of the various disciplines. It is also clear how there has been a sharp division between natural sciences and sociology since the latter's inception, with the aim of affirming its autonomy as a science. This fracture inaugurates at the outset of the last century and lasts until the 1970s, when the issue of "the crisis of sociology" began to arise. A lively debate commences, particularly in the American world, manifested also by ample coverage on scientific journals. Aside from such occasions, often controversial and sterile, many reflections range far and wide within the sociological epistemological framework and the interactions between sociology and other sciences. These considerations often translate into a true defence of this science (see, for example, C.A. Ellwood and P.A. Sorokin). These debates, however, do not question the autonomy of sociology, as this would be to misinterpret the issue. The discussion is not about the autonomy of a discipline, but about the need to solve a human problem with responses at different levels (De Giacinto 1965), all of which concern the solution to a single object of investigation.

3.1   The Interplay Between Sociology and Psychology in the American Debate of the Last Century

Starting from the above remarks, we will now try to outline the evolution undergone by the interplay between sociology and other natural and social sciences, and specifically the interplay between sociology and psychology. Sorokin, for example, is against those ways of "doing" sociology that try to reduce it to pure research technique and to deprive it of the depths of human values and meanings. Sorokin feels that sociology has been betrayed and is firmly opposed to what he calls "quantophreny" and "testomania" (Sorokin 1955) and all the other reductive concepts that he describes as "mechanical" or "robot":

"Hundreds of competent and incompetent psychologists, psychiatrists, anthropologists, sociologists, and educators began to manufacture their own tests and to apply them to hundreds of thousands of human beings, to social groups, and to cultural phenomena. Now and then the manufacturers of intelligence or aptitude tests did not know the ABC's of psychology or sociology; and once in a while they did not have intelligence enough to understand their own incompetence. In spite of these obstacles, multitudes of 'testers' have succeeded in selling their products to their fellow-scholars, educators, governmental agencies, business and labor managers, and to the public at large. [...] we are living in an *age of testocracy*. By their tests of our intelligence, emotional stability, character, aptitude, unconscious drives, and other characteristics of our personality, the testocrats largely decide our vocation and occupation. They play an important role in our promotions or demotions and in our successes and failures in social position, reputation, and influence. They determine our normality or abnormality, our superior intelligence or hopeless stupidity, our loyalty or subversiveness. By all this they are largely responsible for our happiness or despair, and, finally, for our long life or premature death. The enormous influence of tests and testers is primarily due to the supposedly scientific and infallible character of these tests. The testocrats have succeeded in selling their tests as strictly scientific, precise, operational and unerring" (Sorokin 1956: 51–52).

The above quote shows Sorokin's clear opposition to the use of "measurements" in sociology, especially if the latter is reduced to pure profit instead of being a means for improving the living conditions of human beings. Ellwood's reflections are quite similar, albeit less controversial. The scholar states that "Sociology is again in danger of becoming a dead science, of relapsing into a polite amusement of our intellectual classes. This is due in large part to the invasion of the spirit and method of the so-called natural sciences. Whether we like it or not, the question of method is fundamental in the social sciences, for it involves a theory of scientific knowledge" (Ellwood 1931: 15). He argues that while statistics and measurements are important in studying social life, they are not sufficient to explain everything, because the complexity of human life (which is based on the elements of culture) does not follow the causation principle (cause-effect) of physical sciences. The fundamental difference between physics and social sciences is culture. There is nothing similar to culture in the rest of individual existence. Ellwood argues that historical contexts, influences, property, governments, and many other aspects of culture are not stable and fixed, unlike the objects of physical science (LoConto 2011). Ellwood also addresses the issue of the relation of sociology with other sciences. If the relation of sociology to natural sciences is quite clear, its interplay with other social and human sciences, and particularly with psychology, is less so.

"The relation of sociology to the other social sciences, however, is a purely logical relation, and can be fully described only in logical terms. It is the relation of the general to the special. The special social sciences, as the name implies, deal with problems which are relatively specific and concrete, concerning only one section or aspect of the social process. Their generalizations are, therefore, relatively partial and incomplete. Sociology, on the other hand, tries to reach generalizations of a higher order, and to present a general or complete view of the social reality. The social problems which are of a general nature, therefore, that is, those which pertain to the social process as a whole, are left by the special social sciences to sociology" (Ellwood 1907: 316–317).

This, however, does not mean that sociology is the sum of the special social sciences; rather, it appears from all angles as a generalization of social processes. Ellwood claims that sociology is the fundamental science of social life, the foundation of social sciences, as well as their logical completion. To do their job, sociologists must know the results of the special social sciences. Lack of dialogue translates into a dangerous isolation of these special social sciences from sociology and vice versa. Excessive specialization in any social science must be discouraged. The individual life is a unit, and must be studied in all its aspects and from all angles. The close interdependence between various social sciences and sociology must therefore be emphasized.

With regard to psychology, he argues that, besides being the science of individual human nature, it is the basis of all social sciences, since the interactions of individuals in a group are first and foremost psychic ones. "The distinction, then, between sociology and psychology is the same as that between all other sciences—it is fundamentally a distinction of problems. The problems of the psychologist are those of consciousness, of the individual mind, as we commonly say; while the problems of the sociologist are those of the interaction of individuals and the evolution of social organization" (Ellwood 1907: 335). In other words, the distinction between sociology and psychology is the observational point of view. The psychological one is the individual and her experiences (the psychic nature of the individual), while the sociological one is the social group and its organization (the nature of society). Under this perspective, sociology appears as an application of psychology to the interpretation of social phenomena. Psychology is therefore paramount for the work of sociologists. If it is true, as has been recently declared, that "no one is a psychologist unless he is a biologist", it is even more true that "no one is a sociologist unless he is a psychologist" (Ellwood 1907: 335). Ellwood, while never denying the autonomy of sociology, firmly asserts the need for an interdependence of all social sciences. Similarly, Sorokin also assumes a position of openness towards other sciences. This is never explicit, it's the watermark in many of his works on the study of personality. Nevertheless, he continues to emphasize, if not to "shout", the autonomy and object of study of sociology:

> "Of all the other sciences that of prime importance in relation to sociology is psychology. The indisputable fact that human interaction is essentially psychic, involving the exchange of feeling, ideas, wishes, etc., led many sociologists to assert that sociology is nothing more than collective psychology, and that it is based solely on psychology. The claims of this so-called psychological school would have been valid, were it true that psychology deals with the phenomena of human *inter*action. Such an assertion, however, would belie the facts. The focal point of the psychologist's interest is not phenomena of an *inter*mental or *inter*human nature, but the content, structure, and processes of the *individual* consciousness or behavior. The relation of psychology to sociology in somewhat similar to that of anatomy, physiology and morphology to ecology just as, the first three deal only with the intra-organic, leaving to ecology the treatment of the interorganic, so is the sociology limited to the study of the intra-individual like anatomy, psychology breaks up the individual consciousness into such elements as will, feeling, emotion, instinct; like psychology, it studies the processes within the limits of the individual consciousness; and like the morphology it classifies them according to types; it never goes beyond the realm of the individual, the intermental processes of communication between individuals and their mutual actions and

interactions being of interest to the pure psychologist. [...] Our examination into the nature of psychology and of the problems with which it deals revealed conclusively that the realms of the phenomena treated by psychology and sociology are absolutely distinct, affording no basis for their identification" (Sorokin n.d.: Chap. II, pp. 9–11).

In light of his extreme advocacy of sociology as a science (and as an autonomous science), Sorokin's position on the interplay between this and other sciences is exactly what we would have expected. If this is his position on what was defined as *individual psychology*, his idea on *collective or social psychology* is not different. The latter is defined as the discipline that seeks to understand the way in which the real, imagined or implicit presence of others influences the thought, feelings, and behaviours of individuals (Allport 1968). In fact, "If social or collective psychology is the study of the fundamental forms of human interaction and their results, then its subject matter coincides with that of sociology, and which name we use is of slight importance. [...], then this science is simply a section of sociology, which remains the science of the fundamental forms of interaction" (Sorokin n.d.: Chap. II, p. 12). In other words, what Sorokin wishes to emphasize is that sociology does not stop at the study of interaction processes (aspects that can be called mechanical), but it analyses the products and meanings of such processes.

Small (1906), for example, clarifies that the distinctive element of sociology is the mapping of all human experience as a functional process in which the elements of this experience have their own meanings. According to this author, the relation between sociology and psychology is too obvious to give space to the discussion: psychologists and sociologists have so much in common that none of them can afford to leave the other out of their own perspective for too long.

This, in general terms, was the debate at the beginning of the last century (especially in America) focused on the relation between sociology and psychology. A debate often based on dialectics entailing also polemics about the supremacy of one discipline over the other. The fundamental questions that were raised concerned the methods and, above all, the object of study of these two disciplines (aspect that marked the greatest differences). In both cases, while the autonomy of the two sciences always emerges, one cannot but understand that even critical scholars or those who adopted an isolated position choose, substantially or latently, to open up to the interaction between the two sciences.

3.2   Modes of Interplay Between Sociology and Psychology

Historically and in general terms, it can be argued that the interplay between the two sciences has evolved and developed in different ways, also due to the scholars' distinct geographical contexts.

One mode has been characterized by the influence of psychology—very often psychoanalysis (from Freud to Adler, up to Jung)—on the conceptual and theoretical framework of sociology, as well as by the sociological works written by some psychoanalysts. An example in this sense is the evolution of Durkheim's concept of collective representation (1898). This notion, suitable for describing traditional societies, could not survive in the face of a social context so deeply changed with the advent of modern society. This required a major rethinking, also due to having overcome the dominant positivism characterizing most of the social sciences since the beginning of the twentieth century. This change of perspective profoundly affected all disciplines, including sociology, but especially social psychology, that had to redefine the Durkheimian concept of collective representation, transcending its static and coercive character. Thanks to Moscovici's monumental theoretical work (1961), since his important research on the diffusion of psychoanalysis in France, the adjective "collective" is replaced by "social". This highlights the reductive character with which Durkheim's sociology had framed the problem, and introduces the new element of conceiving representations starting from the structure. Social representations are no longer understood as a concept, but rather as a phenomenon.

The other mode, instead, concerns the application of psychology—in its individual (psychoanalysis) and social variants—to the sociological investigation carried out by sociologists. This phase is characterized by the presence of three schools. (1) The first can be traced back to the Frankfurt School (Horkheimer, Adorno, Marcuse, Fromm and later Habermas) that promoted a series of researches basing them on the integration of the analysis of capitalism and that of the individual. This was not only a criticism of society, but also of its culture and the individual. In particular, the Frankfurt School grounds all its reflections on two main assumptions: (a) the ideas of individuals are a product of the societies in which they live and therefore it is impossible to achieve an objective knowledge that is not influenced by the conceptual models of the society of reference; (b) intellectuals, in carrying out their duties, should not seek to be objective, rather they should adopt a critical attitude towards society. The aim is to guide the individuals towards awareness of what they should do to generate social change; (2) The second "school" is identified—and coincides with—Talcott Parsons' works. Parsons used the theoretical conceptual framework of psychology within the sociological theoretical framework, with several methodological precautions aimed at minimizing any risk of psychological reductionism and the arbitrary transposition of psychoanalytic categories. In particular, the American scholar uses the theoretical frameworks of psychology by applying them to his theory of action and social system (Parsons 1937, 1951, 1964). Concerning the relation between the two disciplines, he states:

> "sociology is clearly concerned with the observation and analysis of human social behavior, that is, the interaction of pluralities of human beings, the forms their relationships take, and a variety of the conditions and determinants of these forms and of changes in them. The psychologist is traditionally concerned with the behavior of "the individual" though a very large part of the behavior of individuals occurs in relationships with other individuals. Sometimes there is of course even more overlap as when "social psychologists" concern themselves with the behavior of crowds, the formation of public opinion, and the like. [...] From the present point of view the focus of sociological theory is held to be on certain aspects of the structure of and processes in social systems. A social system in turn I define as the system constituted by the interaction of a plurality of human beings, directly or indirectly, with each other. Psychology, on the other hand, I hold to be concerned first with certain elementary processes of behavior, like learning and cognition, which, however

much they may be concretely involved in social interaction, can be isolated from its processes for special study, and secondly with the organization of the components of behavior to constitute the personality of the individual as a system, the system of behavior of a single specific living organism. This way of defining the relations of the two theoretical disciplines has certain implications which should be made explicit. Their common reference is behavior (Parsons 1954: 68–69);

(3) The third school can be ascribed to those scholars who, since the 1940s, gathered around the Tavistock Institute in London, dealing mainly with industrial sociology, sociology of organization, and sociology of work. They not only used psychological theories, but also applied some psychoanalytic therapeutic practices to introduce new research and intervention techniques in the field of work organization.

A last mode, less incisive than the previous ones, concerns the influence of sociological theory on psychological theories. In fact: "the direction of strongest influence has run from psychology to sociology, rather than the reverse. This is in part because sociologists generally devote their efforts to identifying *which* social phenomena have effects on individuals while psychologists generally specialize in identifying *the mechanisms or processes through which* social phenomena have their effects on individuals" (Thoits 1995: 1231). It is therefore clear how the influence of sociology on psychology is of slight importance or even non-existent, unlike the opposite, which seems to be obvious despite the presence of several points of reciprocity.

4   Towards a New Scientific Awareness: Transdisciplinarity

Drawing a first conclusion, we could say—following Levinson (1964)—that the relation between sociology and psychology until the 1940s can be summarized by the expression "'separate but equal'. As we know so well from the study of race relations, those who emphasize the boundaries between two groups are usually more concerned with separateness than equality. In their efforts to define and legitimize a new disciplinary identity, the early sociological theorists laid stress on sociology's separateness from psychology" (Levinson 1964: 78). The passing decades have not completely changed this situation, but some of the harshest polemics have certainly been dampened and there is more collaboration between scholars from the two disciplines. According to Levinson, researchers who join forces to build a sort of interdisciplinary bridge are often victims of an identity-related anxiety. They fear that their professional skills may be overshadowed. There is no lack of collaborative efforts, and representatives of single disciplines demonstrate an increasing ability to assimilate the concepts and methods of others. The increase in interdisciplinary research has promoted a new way of understanding research in the field of social and psychological sciences, making a significant contribution through the development of: (1) a more complex conception of human individual personality; (2) a more complex conception of the sociocultural environment of the individual; and (3) concepts that represent the interactions between the previous two points.

More recently, an example of such interdisciplinarity can be found in cultural psychology (Shweder 1990; Valsiner 2014), an emerging field of psychology since the 1990s that today finds its full affirmation. "Cultural psychology is the study of the way cultural traditions and social practices regulate, express, transform, and permute the human psyche, resulting less in psychic unity for humankind than in ethnic divergences in mind, self, and emotion. Cultural psychology is the study of the ways subject and object, self and other, psyche and culture, person and context, figure and ground, practitioner and practice live together, require each other, and dynamically, dialectically, and jointly make each other up" (Shweder 1990: 1). In Valsiner's words, cultural psychology captures the complexity and dynamics of experiences: "With the emergence of cultural psychology since the 1980s there may be a new chance to capture the complex and dynamic phenomena of human experiencing. But for that to happen, many of the existing ways in which psychology creates its knowledge needs a constructive overhaul" (Valsiner 2014: 7). As Marsico (2015) says, cultural psychology is a science of development by its very nature. It assumes that all human beings, as well as their forms of organization (groups, communities, institutions), develop dynamic systems that are constantly engaged in searching for the "new". Thus, the focus of investigation of cultural psychology is the analysis of the situations in which this "new" emerges.

In light of all this, rather than drawing further conclusions, we will express a wish for the scientific future of social sciences. The better solution to untie the knot of the interplay between sociology and psychology is to allow greater collaboration between sociologists and psychologists by sharing, within the same theoretical framework, different theories, methodologies, and methods of analysis. Progress in this direction will lead to changing the traditional 'separate but equal' situation and the new budding relationship will offer greater promise for the development of humanity. And this is all the more necessary as the multidimensional crisis of society (economic, social, cultural, etc.) worsens, and attempts to define and launch new policies (economic, welfare, citizenship, etc.) could not prevent the depletion of legal protections and of the social fabric, that needs to be reconstructed. Sorokin ends his career as a scholar precisely trying to make this clear to his colleagues, but they opposed him for this very reason.

In this reconstruction process lies the knowledge of human and social sciences, that must pay attention to all aspects of transformation of society and not just to specific areas. The researcher's actions cannot be exclusively technical: knowledge of human and social sciences, and particularly sociological knowledge, can explain and understand the complexity of the problems and situations that individuals experience in everyday life. This is because, if order characterized traditional societies, disorder characterizes contemporary societies, and this forces scholars to redefine paradigms and methods. Knowledge must be configured as an experience of exchange resulting from "confrontations" and "conflicts" between disciplines beyond any real or virtual boundary that delimits the "spaces of movement" (D'Angelo and Mangone 2016). The perspective of transdisciplinary cutting becomes of paramount importance precisely on the basis of the latter statement. This is not a super-discipline, but as a new approach that drives knowledge to a higher stage "which would not merely achieve interactions or reciprocities between specialized researches, but would place these links within a whole system without stable boundaries between disciplines" (Piaget 1972: 170).[7]

The aim must be to concretise cooperation between the disciplines. The multidimensionality of daily life issues and the rapid succession of societal transformations drives us to recompose the different points of view and perspectives of the various disciplines. It is necessary to open a dialogue that overcomes the "formal" disciplinary and terminological barriers. For this reason, Lazlo said in an interview:

"Disciplines in science are artifacts; they are artificial. They are often necessary, but not always a satisfactory limitation on the number of observations and the number of facts that one takes into account. There are no boundaries in nature that correspond one to one with the boundaries of disciplines. For example life is not necessarily limited to biology, it's also obviously evident in sociology and psychology. It also appears in the cosmos. The way we can think about evolution is not limited to one kind of system. It appears from the big bang onwards all the way up to the evolution of consciousness, the evolution of the whole cosmos at the same time. So disciplines are a necessary self-restriction in science, but they should be considered as permeable, as transferrable and expandable boundaries that one keeps to as long as they are useful. When we can get over these boundaries, then it's an improvement when you manage to overcome them" (Marturano 2013).

Only from permeable and flexible boundaries between disciplines it is possible to attain a knowledge that is "free" from positivism and that can try to provide answers to societal emergencies. This means going "beyond the disciplines" (Nicolescu 1985) and recognizing them as "other disciplines" rather than inferior or superior ones. Therefore, knowledge should become reflexive in promoting the building of relationships within the living environments of individuals and between individuals—acknowledging the autonomy of individual disciplines, but abandoning the excess of self-referentiality that encloses them within their own frameworks and paradigms.

References

Alexander, J. C. (1982). *Theoretical logic in sociology. Volume 1: Positivism, presuppositions and current controversies*. London: Routledge and Kegan Paul.
Allport, G. W. (1968). *The person in psychology: Selected essays*. Boston: Beacon.
Collins, R. (1988). *Theoretical sociology*. Orlando: Harcourt Brace Jovanovich.
Comte, A. (1830). *Cours de philosophie positive* (Vol. 6). Paris: Bachelier.
Coser, L. A. (1971). *Masters of sociological thought*. New York: Harcourt Brace Jovanovich.
D'Angelo, G., & Mangone, E. (2016). Beyond the disciplinary borders: A new challenge. *Journal of Mediterranean Knowledge, 1*(1), 3–9.
De Giacinto, S. (1965). In occasione della traduzione italiana di fads and foibles in modern sociology di P.A. Sorokin. *Studi di Sociologia, 3*(4), 342–350.
Durkheim, É. (1893). *De la division du travail social*. Paris: Alcan.
Durkheim, É. (1895). *Les règles de la méthode sociologique*. Paris: Alcan.

---

[7] The French translation of this quote and any mistakes made in it are the responsibility of the author of the book.

Durkheim, É. (1897). *Le Suicide: Étude de sociologie*. Paris: Alcan.

Durkheim, É. (1898). Représentations individuelles et représentations collectives. *Revue de Métaphysique et de Morale, VI*, 273–302.

Ellwood, C. A. (1907). Sociology: Its problems and its relation. *American Journal of Sociology, 13*(3), 300–348.

Ellwood, C. A. (1931). Scientific method in sociology. *Social Forces, 10*, 15–21.

Giddens, A. (1979). *Central problems in social theory: Action, structure and contradiction in social analysis*. London: McMillan.

Goldthorpe, J. H. (2000). *On sociology. Numbers, narrative, and the integration of research and theory*. Oxford: Oxford University Press.

Gouldner, A. W. (1970). *The coming crisis of western sociology*. New York: Basic Book.

Habermas, J. (1984). *The theory of communicative action: Volume 1: Reason and rationalization of society*. Boston: Bacon.

Habermas, J. (1987). *The theory of communicative action: Volume 2: Lifeworld and system: a critique of rationalism reason*. Boston: Bacon.

Kuhn, T. (1962). *The structure of scientific revolution*. Chicago: The University of Chicago Press.

Levinson, D. J. (1964). Toward a new social psychology: The convergence of sociology and psychology. *Merrill-Palmer Quarterly of Behavior and Development, 10*(1), 77–88.

LoConto, D. G. (2011). Charles A. Ellwood and the end of sociology. *The American Sociologist, 42*, 112–128.

Manicas, P. T. (1987). *A history and philosophy of the social sciences*. Oxford: Basil Blackwell.

Marturano, A. (2013). A theory of everything—Ervin Laszlo and Antonio Marturano. *Integral Leadership Review, 13*, 1. Retrieved June 20, 2015, from http://integralleadershipreview.com/8020-fresh-perspective-a-theory-of-everything-%C2%AD-ervin-laszlo-and-antonio-marturano/.

Marsico, G. (2015). Striving for the new: Cultural psychology as a developmental science. *Culture & Psychology, 21*(4), 445–454.

Merton, R. K. (1968). *Social theory and social structure*. New York: The Free Press.

Morin, E. (1984). *Sociologie*. Paris: Fayard.

Moscovici, S. (1961). *La Psychanalyse: Son image et son public*. Paris: PUF.

Nicolescu, B. (1985). *Nous, la particule et le monde*. Paris: Le Mail.

Nisbet, R. A. (1977). *The sociological tradition*. New York: Basic Book.

Parsons, T. (1937). *The structure of social action*. New York: McGraw-Hill.

Parsons, T. (1951). *The social system*. New York: Glencoe.

Parsons, T. (1954). Psychology and sociology. In J. Gillin (Ed.), *For a science of social man: Convergences in anthropology, psychology, and sociology* (pp. 67–101). New York: Macmillan.

Parsons, T. (1964). *Social structure and personality*. New York: Glencoe.

Piaget, J. (1972). L'épistémologie des relations interdisciplinaires. In OCDE (ed.), *L'interdisciplinarité: problèmes d'enseignement et de recherche dans les universités*. Paris: OCDE. Retrieved June 20, 2015, from http://www.fondationjeanpiaget.ch/fjp/site/textes/VE/jp72_epist_relat_interdis.pdf.

Shweder, R. A. (1990). Cultural psychology—What is it? In J. W. Stigler, R. A. Shweder, & G. Herdt (Eds.), *Cultural psychology* (pp. 1–43). Cambridge: Cambridge University Press.

Small, A. E. (1906). The relation between sociology and other sciences. *American Journal of Sociology, 12*(1), 11–31.

Sorokin, P.A. (n.d.). *The nature of sociology and its relation to other sciences*. University of Saskatchewan: University Archives & Special Collections, P.A. Sorokin funds, MG449, I, A, 3.

Sorokin, P. A. (1931). Sociology as a science. *Social Forces, 10*(1), 21–27.

Sorokin, P. A. (1937). *Social & cultural dynamics: Vol. I: Fluctuation of forms of art*. New York: American Book.

Sorokin, P. A. (1941). Declaration of independence of the social sciences. *Social Sciences, 16*(3), 221–229.

Sorokin, P. A. (1955). Testomania. *Harvard Educational Review, XXV*(4), 199–213.

Sorokin, P. A. (1956). *Fads and foibles in modern sociology and related sciences*. Chicago: Henry Regnery.

Sorokin, P. A. (1962). *Society, culture, and personality: Their structure and dynamics, a system of general sociology*. New York: Cooper Square.

Sorokin, P. A. (1966). *Sociological theories of today*. New York/London: Harper & Row.

Sorokin, P. A. (1998). The boundaries and subject matter of sociology. In B. V. Johnston (Ed.), *On the practice of sociology* (pp. 59–70). Chicago: University of Chicago Press.

Thoits, P. A. (1995). Social psychology: The interplay between sociology and psychology source. *Social Forces, 73*(4), 1231–1243.

Valsiner, J. (2014). *An invitation to cultural psychology*. London: Sage.

Wallace, R. A., & Wolf, A. (1991). *Contemporary sociological theory: Continuing the classical tradition*. Englewood Cliffs: Prentice Hall.

Weber, M. (1947). *The theory of social and economic organization*. London: Collier-Macmillan.

Weber, M. (1978). *Economy and society: An outline of interpretive sociology*. Berkeley: University of California Press.

Zoll, R. (2000). *Was ist solidarität heute?* Frankfurt am Main: Suhrkamp Verlag.

Chapter 3
Integral Theory of Knowledge

1 The Dyad Knowledge/Reality

The methodology identified by Sorokin for the social sciences studying the "super-organic" world (sociocultural phenomena) is based on the idea that sociocultural phenomena differ from inorganic and organic ones. This methodology, called Integralism or integral theory of sociology—clearly described by Sorokin in *Integralism is My Philosophy* (1958)—is part of his broader theory of the sociology of knowledge and of integration of the systems of truth. This distinctive element of Sorokin's Integralism makes it necessary—before diving into its difficult analysis and understanding—to explore the relationship between knowledge and reality.

But even before that, we should clarify that this—despite the name—is not a philosophy, and he is not a philosopher. He proposes a theory of knowledge that is a general idea, a model of individuals and society in their constant evolution, representing his theoretical premise for sociological analysis rather than a philosophical system. He is not a philosopher because his *forma mentis*, as well as the logical procedures, verifications, and methods he adopted, remain closely related to the sociological perspective and its object of study.

Knowledge and reality are bound in an indissoluble couple and the relationship between them has always been studied, right from classical philosophy (search for truth):

"the sociology of knowledge must analyse the process in which this occurs. The key terms in these contentions are 'reality' and 'knowledge', terms that are not only current in everyday speech, but that have behind them a long history of philosophical inquiry. We need not enter here into a discussion of the semantic intricacies of either the everyday or the philosophical usage of these terms. It will be enough, for our purposes, to define 'reality' as a quality appertaining to phenomena that we recognize as having a being independent of our own volition (we cannot 'wish them away'), and to define 'knowledge' as the certainty that phenomena are real and that they possess specific characteristics. It is in this (admittedly simplistic) sense that the terms have relevance both to the man in the street and to the philosopher. The man in the street inhabits a world that is 'real' to him, albeit in different degrees, and he 'knows', with different degrees of confidence, that this world possesses such

© The Author(s) 2018

E. Mangone, *Social and Cultural Dynamics*, SpringerBriefs in Psychology,
https://doi.org/10.1007/978-3-319-68309-6_3

and such characteristics. The philosopher, of course, will raise questions about the ultimate status of both this 'reality' and this 'knowledge'" (Berger and Luckmann 1966: 13).

One feature of this definition refers to the fact that reality, or groups of reality, belong to particular social contexts (social relativity), and it is precisely this peculiarity that originally justifies the sociologists' curiosity for both reality and knowledge. From this point, the scholars Berger and Luckmann take a further step by considering the differences in "knowledge" in the various societies as an indisputable fact. The sociology of knowledge should not only study the empirical variety of knowledge, but also the processes through which every knowledge structure is socially established as a reality. The sociology of knowledge must, therefore, deal with what individuals "know" as "reality" in everyday life. It has to study common sense rather than theoretical knowledge, since that is the first form of knowledge influencing and guiding the daily action of individuals. Thus, the "world of everyday life" becomes the metaphorical "place" in which to carry out the analysis. Within it, people express intentionality-oriented attitudes towards objects that, in turn, through differentiation, appear to the consciousness as constituent of different "spheres of reality". The reality of everyday life is: (a) "ordered", because the phenomena are predisposed independently from their perception and are imposed on the latter; (b) "objectified", because its perceived order of objects precedes the presence on the scene of individuals.

The need to deal analytically with the "gnoseological problem", namely the problem of the value of knowledge (*gnòsis*), is linked to the fact that the primary function of knowledge is the construction of meanings and social reality (Mangone 2012, 2015). The concept of "knowledge" thus defines those sets of meanings and interpretations that the individual processes and ascribes to the data and information she gathers from the context in which she lives. Knowledge is the result of a constructive process of reorganization, elaboration, representation, and interpretation of the information. These processes involve, at the same time, psychological aspects (perceptions, emotions, cognitions) and social, cultural, and historical ones, that allow their transformation into models and representations. In a nutshell, to know means: (a) to participate in the construction of meanings of social and cultural reality in order to transform it in symbolic representation; (b) to attribute "sense" and "meaning" to events, objects or individuals on the basis of knowledge, expectations and assumptions; (c) to develop in a complex and dynamic way the information that individuals can gather and turn them into knowledge.

The ways in which people interpret and perceive the world is a traditional element of Western thought. The perceptual processes (primary type of information) that Luhmann (1983) describes as the psychic acquisition of information can also be described as adaptation to change or as the beginning of a change. The relation between how human beings think of their social action and how they perceive it (conceptual process and perceptual process) is a complex process of interaction and acquisition of knowledge. In this process, communication (and especially language) is a form of objectivation of human expression. We cannot help making a brief reference to the founder of modern linguistics. Ferdinand de Saussure (1971) claimed that the sign (linguistic sign—semiotic event in the *stricto sensu*) is a complex event that cannot exist in the absence of one of the two elements that constitute it. The *concept* (signified), that can only be defined within a signification process, is a psychic representation of the "thing", not an act of consciousness nor a reality (Barthes 1964). And the *image* (significant)—acoustic or graphic—that is a creation originated from the selection of certain characteristics of the acoustic or graphic sequence. And since the various forms of communication, including language, are the basis of the cognitive processes, it follows that the two elements that make up the sign (significant and signified) cannot be considered separately. Knowledge is in fact the product of communication and it represents the relationship between the intended meaning of a situation (or text) and the part of it that the individual interprets on the basis of her own experience and purposes.

On the latter aspects, according to Moscovici (1984), society is divided into two universes of knowledge and each of them forms a reality. The *reified universe* is characterized by scientific causality (a posteriori explanation), as it attributes causes according to theories or explanatory models that are legitimized and shared by the scientific community. The *consensual universe* is characterized by social causality, where both effects and causes are directly related to the social representations of individuals (identification and recognition of the situational context). This argument is quite similar to the one made by Schütz (1946) in his essay on the social distribution of knowledge. He emphasizes the importance of the processes and mechanisms involved in passing from an "expert" knowledge to a knowledge related to the experiences of the "man on the street" in his daily life. In recent times, the philosopher Lévy (1994), who has long considered the relationship between knowledge and everyday life and the consequences of the diffusion of digital broadcasting networks, suggests a possible solution to this "compartmentalized" knowledge. The way out is represented by what he calls *collective intelligence*, understood as an intelligence disseminated everywhere, continuously valued, coordinated in real time, leading to an effective mobilization of skills. Every interaction with objects or with humans, every act of communication, implies a transmission of skills and knowledge, thus an exchange that becomes a process of integration of the differences—understood as collective wealth in which everyone is recognized—without limitations in the learning paths of each individual or prejudices on "expert" and "profane" knowledge.

Knowledge is therefore the ultimate goal of humanity. Many debates have been kindled about this process, starting with philosophy as the first human-related science, followed by other social sciences, including the sociology of knowledge (*Wissenssoziologie*). The sociology of knowledge is a continuous confrontation between different ideas of the world that materialize in a specific space and time: it is the study of the connections between categories of thought, forms of knowledge, and social structures. Its study approaches[1] are many, but in the present work we

---

[1] The main representative approaches in the history of sociology are the following: Durkheim's proto-functionalism and its more advanced form as structural-functionalism in Merton's version; Marx's theory of conflict and the Fankfurt School's critic of society; Mannheim's and Sorokin's *Wissenssoziologie*; and, finally, the phenomenological approach (Scheler, Schütz, Berger, and Luckmann).

will consider only Sorokin's sociology of knowledge, because he tried to synthesize together the main sociological approaches that had focused on the relationship between thought systems and social systems.

2   Sorokin's Sociology of Knowledge

According to Sorokin,

"One of the main tasks of so-called sociology of knowledge (*Wissensoziologie*) is a study of the factors which condition the essential contents, configurations, and transformations of the mental life of an individual or of a group: [...]. The sociology of knowledge or, more exactly, the sociology of mental life tries to answer the basic questions of how and why the mental life of any given individual or of a group happens to be such as it is and how and why it often changes in the course of the individual's or group's life, and why the mental life of various persons or collectivities is often quite different. The sociology of mental life endeavours to elucidate these problems through a study of the mentalities of vast cultures and societies (macrosociology of mental life) and through that of the mental life of a given individual (microsociology of mental life)" (Sorokin 1963a: 3).

The above statement is based on two assumptions: one regards the individual as a three-dimensional creature (body, mind, and soul), and the other states that it is important to define the concept of human culture in a very broad sense. Sorokin, in *Social & Cultural Dynamics*, states that "*In the broadest sense it may mean the sum total of everything which is created or modified by the conscious or unconscious activity of two or more individuals interacting with one another or conditioning one another's behavior*" (1957: 2). However, a few pages later, he specifies that the many interrelations between the various elements of the culture can be reduced to four basic types: (1) *Spatial or Mechanical Adjacency*, whereby the elements are held together mechanically or accidentally (for example, a mixed garbage heap); (2) *Association Due to an External Factor*, whereby the cultural elements are adjacent to each other but have no functional or logical connection (for example, skis, grappa, and chimney in winter); (3) *Causal or Functional Integration*, whereby the elements are connected in a functional way (for example, the components of a car when assembled); and finally, (4) *Internal or Logico-meaningful Unity*, which represents the highest possible form of integration. Without it, Michelangelo's Pietà would be a piece of marble like any other, and Vivaldi's Four Seasons a series of dots on paper sheets. These types of integration also distinguish "systems" from "congeries". These latter are just any conglomeration of cultural elements (objects, traits, values, ideas) in a given area of social and physical space, with spatial or mechanical concurrence as the only bond of union.

From this, it follows that the cultural systems are characterized by logical integration (*Internal or Logico-meaningful Unity*) that can be studied in two ways:

"The elements of thought and meaning which lie at the base of any logically integrated system of culture may be considered under two aspects: the *internal* and the *external*. The first belongs to the realm of inner experience, either in its unorganized form of unintegrated images, ideas, volitions, feelings, and emotions; or in its organized form of systems of thought woven out of these elements of the inner experience. This is the realm of mind, value, meaning. For the sake of brevity we shall refer to it by the term "mentality of culture" (or "culture mentality"). The second is composed of inorganic and organic phenomena: objects, events, and processes, which incarnate, or incorporate, or realize, or externalize, the internal experience" (Sorokin 1957: 20).

In the light of the peculiarities of the two aspects, Sorokin raises a question that we can formulate as follows: Is it possible to appropriately grasp the inner aspect of a given culture?

Obviously, the answer to this question depends on what is believed to be the true mentality or the real meaning embedded in a set of external signs. A real meaning can be the one offered by psychological interpretation: for example, attributing meaning to the Divine Comedy based on the meanings in Dante's head. According to this view, the correct reading of the inner aspects of culture (the real meaning) coincides with those envisaged by its creators. In the above example, the real meaning is the one that Dante had attributed to his work. But psychological interpretation is not enough to read the whole sociocultural phenomenon. Sorokin suggests another reading for the internal aspects of cultural phenomena called *sociologico-phenomenological:*

"What are the essentials of the sociologico-phenomenological reading? First there is a causal-functional reading of culture which aims to discover the *causal-functional relationships* between the component parts of a cultural value or complex. [...] A second form of the sociologico-phenomenological interpretation of the internal aspect of culture is the *logical reading*. [...] The necessity for the logical interpretation follows also from the limitations of the psychological reading. The point is, in other words, that most of the cultural phenomena represent the results of the activities of many individuals and groups, whose purposes and meanings may be different from one another, often opposite. [...] We may sum up this discussion with the statement that a proper logical reading of cultural phenomena requires: first, the application of the canon of deductive and inductive logic; second, the realization of the possibility that the major premises of various cultures may differ; third, the assumption of an impartial position in regard to the validity or invalidity of the major premises" (Sorokin 1957: 21–23).

Sorokin's sociology, and consequently his integral theory of sociology, stems and develops from his theory of cyclical movements of systems (Ideational, Sensate, and Idealistic, respectively),[2] which are produced by the transformations of the

---

[2] The Ideational refers to theological science: its object is mainly the supersensory, and superrational "subjects" and "realities": the sensory and empirical phenomena are studied only incidentally and even then not for their own sake but merely as "visible signs of the invisible world" as symbols of the supersensory reality, and its validity criterion refers to the Sacred Scriptures, where logical reasoning is utterly superfluous and it is recognized only when sensory perception does not contradict the Scriptures. In the Sensate system, sensory reality is prevalent and the relationship between human being and society is instrumental: the world of the sensory perception, like the phenomena studied in the natural sciences, and its validity criterion is the reference to the testimony of the organs of senses, supplemented by the logical reasoning, especially in the form of the mathematical reasoning. The last system, the Idealistic system is mixed, it subsumes elements of the Ideational and Sensate system and its object is partly supersensory, partly sensory-empirical. Each for its own sake, but the value of the knowledge about the sensory phenomena is subordinated

mental bases of human beings and groups. Sorokin's interest, therefore, focuses on the study of the dynamics of the historical transformations of sociocultural systems, starting precisely from the typology of systems existing in the history of humanity that characterize certain values and particular forms of knowledge.

Since the character of a culture is determined by its internal aspect—what has been called *mentality*—one must start the study of sociocultural systems from the basic premise of their mentality: (1) *the nature of reality*, characterized by the fact that the same set of material objects that make up an environment is not perceived and interpreted in the same way by every individual; (2) *the nature of the needs and ends to be satisfied*, which may be merely *carnal* or *sensual* (such as hunger or sex), or *spiritual* (the salvation of one's soul or morality), or *mixed* or *carnalspiritual* (scientific or artistic affirmation for oneself and for humanity); (3) *the extent to which these needs and ends are to be satisfied*, namely the establishment of a measure in which these needs and ends must be met (hunger can be satisfied with a piece of bread or a number of courses); (4) *the methods of satisfaction*, which follow three directions: changing the environment to satisfy the need; acting on the individual to change the hierarchy of needs; a combination of the two.

Based on these crucial premises, Sorokin builds his typology (Ideational, Sensate, and Idealistic), that allows him to develop a dynamic theory of culture in relation to society. From his analysis, it emerges that the three systems follow one another through history, transforming and developing until their exhaustion and replacement by another system. Sorokin's contribution to the sociology of knowledge is linked to the idea that cognitive forms are influenced not only by mental categories but also by sociocultural ones (cultural mentality), which in turn are influenced by collective life. Sorokin's merit, then, was to show the great variety of interrelations that are established among the different elements forming the social and cultural system. This consideration leads Sorokin to affirm in *Contemporary Sociological Theory* (1928) the inadequacy of the principles and methods of natural sciences to study social phenomena, because in this way scholars ignored several specific aspects of sociocultural phenomena that belong only to the world of human beings. Throughout his thinking, he has always highlighted the need for the development of an integral study of society (integralism or integral theory of knowledge), because the true nature of human interaction cannot be understood only by describing behaviours (behaviourism) but, rather, by penetrating deep into the inner meaning of interactions. This understanding can be acquired through a continuous search combining these aspects with the external ones of the actions, desires, and aspirations of individuals.

---

to that of the supersensory "realities". The total system of knowledge here incorporates, usually in the form of idealistically rationalistic Philosophy, but the ultimate reality is thought of as knowable. The exposition of the truth is dialectic and deductive. In this system logical reasoning prevails, without disregarding references to sensory experience, in fact, it integrates the three forms of truth (sense, reason, and faith).

3   The Integralism or Integral Theory of Sociology

The need to move towards an integrated system of knowledge in the study of socio-cultural phenomena is inherent in the very complexity of these phenomena. Following the logic of integration (without excluding the various methods) means to hold together theory, research, and operability. This attributes to sociology and social sciences a renewed role as sciences able to provide not the solution, but rather the feasible paths for the improvement of social problems. In turn, this translates into a continuum of interdependencies ranging from theory to operability, to usability. All this is achieved by means of research-action, that becomes indispensable for the acquisition of a knowledge able to read social phenomena with the participation of the actors themselves (the theoretical premises translate into concrete actions). The general theme of the relationship between theory and practice has gone hand in hand with sociology since its positivist stage. However, the prevalent empirical content of sociological knowledge does not solve the problem of its usability, nor it clarifies the ambivalent role of the researcher (actor and observer at the same time of the phenomenon investigated). These aspects prompt the researcher to wonder what is the most appropriate methodology that relates theory and action to sociocultural phenomena. Research is a tool to expand the ability to *describe* the phenomenon, by increasing the knowledge that leads to its *explanation* and *understanding*, and then to its *prediction*. These levels are neither sequential nor separate (Homans 1967), but rather constitute a unique set that translates into the integration between theory, research and operability, and between the various social sciences. Methodological integration does not aim at confusing knowledge and action. It builds a bridge between these levels of analysis so that they enrich, rather than depleting, each other.

The complexity of sociocultural phenomena is not new: this aspect has been known since the birth of sociology (studying the whole and not the parts). The complexity of sociocultural phenomena is determined by the fact that interactions occur between three interdependent and inseparable aspects, and for this reason the analysis of the generic common properties of sociocultural phenomena should not be considered as the "simplest units". According to Sorokin, these aspects are:

"(1) *personality* as the subject of interaction; (2) *society* as the totality of interacting personalities, with their sociocultural relationships and processes; and (3) *culture* as the totality of the meanings, values, and norms possessed by the interacting persons and the totality of the vehicles which objectify, socialize, and convey these meanings. [...] None of the members of this indivisible trinity (personality, society, and culture) can exist without the other two. There is no personality as a *socius*, bearer, creator, and user of meanings, values, and norms without a corresponding culture and society; only an isolated biological organism can exist in their absence. Similarly, there is no superorganic society without interacting personalities and a culture; and there is no living culture without interacting personalities and a society. Hence none of these phenomena can be properly investigated without considering the other members of the trinity" (Sorokin 1962: 63–64).

If these are the three inseparable and interdependent aspects, sociocultural phenomena and the processes of meaningful human interaction consist of three components: (1) the human beings (both individuals and groups) that create, use, communicate, and exchange values through means (vehicles); (2) the *meanings* (values and behavioural rules) that can be defined as those meanings that are super-imposed on the biophysical structures of interacting individuals. The meanings can be classified into: (a) cognitive *meanings*, in the strict sense they are the meaning of Aristotle's philosophy, a mathematical formula or the prayer for a believer; (b) meaningful *values*, they are the economic value of a property, the value of religion, science, education, etc., or the value of life or health; (c) *norms*, are the law and eth-ics, or the prescriptions for the construction of houses and cars. These three terms (meaning, value, and norm) can be interchangeably used to denote a general class of significant phenomena superimposed on the biophysical properties of individuals and objects, actions, and events; And finally, (3) *vehicles* (material objects, sensory energies, energies such as sounds, lights, colours, movement, electrical forces, ther-mal forces and other agents) through which intangible meanings and values are "objectified" and "communicated" to others.

As mentioned earlier, sociocultural phenomena are distinguished into two main classes: *aggregates*, that are two or more cultural elements without any relation between them (causal or meaningful); *social or cultural systems* that are made up of phenomena united by a logical or meaningful relation, and by causal relationships of interdependence in which each part (meaning, vehicle, and human being) depends significantly on the other. In turn, the various systems—not the aggregates—are reciprocally related: "Since a system has a triple interdependence of part upon other parts and upon the whole and the whole upon its parts, all parts of a sociocultural system function and change 'in togetherness'" (Sorokin 1956: 268). It follows that each system is characterized by a multiplicity of interdependencies of its parts on the others and on the whole, and of the whole on its parts. Therefore, the three ele-ments (personality, society, and culture) and the three components (human beings, meanings, and vehicles) work together at the same time and undergo simultaneous changes. These are the aspects and elements that constitute a generic sociocultural phenomenon, some of which cannot be considered separately (personality, society, and culture). To understand sociocultural phenomena and their changes in relation to the respective cultural mentalities (Idealistic, Sensate, and Ideational), one must start from understanding the systems of truth and reality characterizing these phe-nomena. To this end, according to Sorokin, it is necessary to integrate the systems of knowledge, since reality in its fullness is inaccessible to any human mind:

> "Integralism is its name. It views the total reality as the infinite X of numberless qualities and quantities: spiritual and material, momentary and eternal, ever-changing and unchange-able, personal and superpersonal, temporal and timeless, spatial and spaceless, one and many, the littlest than the little, and the greatest than the great. In this sense it is the veritable *mysterium tremendum et fascinosum* and the *coincidentia oppositorum* (reconciliation of the opposites). Its highest center is the Creative X that passes all human understanding. In its inexhaustible plenitude the total reality is inaccessible to the finite human mind. However, its main aspects can roughly be grasped by us because we are also its important part. Of its innumerable modes being three forms or differentiations appear to be most important: (1) empirical-sensory, (2) rational-mindful, and (3) supersensory-superrational" (Sorokin 1958: 180).

In other words, Sorokin attempts to address the inadequacy of exclusively empir-ical methods and to create a framework of principles and methods for sociology and social sciences. In part, these principles and methods run parallel between these sciences, and yet they are still profoundly different.

3.1 The Systems of Truth

Sorokin's solution is to create an Integral culture based on Integral truth. This form of knowledge, by integrating the three forms of truth (reason, meaning, and faith), provides a fuller and more valid understanding of reality. According to Johnston (1999), integralism is at the same time epistemology, psychology, sociology, and theory of history. This binds the truth of science, reason, and intuition into a whole. They are the means for obtaining a satisfactory cultural framework, from which to begin understanding life, the world, and the role of humanity. The integral system links the three dimensions of truth so that it is closer to the three-dimensional nature of humanity, as well as to reality. It thus provides a more complete and satisfactory approach to knowledge and understanding. Through integralism one knows the individuals, the things they do, and the cultural system is thus constructed in a more complete way.

In order to fully understand what many defined as "Sorokin's philosophy"—but which is not a philosophy[3]—it is necessary to briefly explain each of the aspects considered as reality systems. The empirical-sensory aspect of reality is perceived through the senses of human beings, or with artificial extensions of them (for exam-ple, the microscope or the telescope). In this case science aims, though not exclu-sively, at reaching a precise understanding of this sensorial aspect. The rational-mindful aspect is primarily understood by our reason, starting from all the various forms that it assumes in all kinds of mathematical and logical thought. Mathematics, logic, and rational philosophy strive to clarify X's form. "Finally, the glimpses of its superrational and supersensory aspect are given to us by truly creative-supersensory and superrational intuition, or 'divine inspiration', or 'flash of enlightenment' of all the creative geniuses" (Sorokin 1958: 180). These "creative geniuses" (essayists, philosophers, writers, artists, and other important creators in the different cultural fields) are the testimony to the whole humanity of the fact that their discoveries or creations have been started and guided by the grace of intuition. They have been the instrument of this creative force that transcends human capabili-ties: every individual is visited by this grace of "the supreme enlightenment". This last form of knowledge was what led to the greatest criticism against Sorokin, and the accusations of practicing not sociology but the "art of prophecy". An intuitive supersensory knowledge is considered a superstition, like a premonition or prophecy,

---

[3] We point out, once again, that Integralism is not a philosophy, because it seeks to maintain focus on the social problems observed through a systematic theoretical approach within a framework that integrates and triangles empirical, sensory, and intuitive understanding.

not a form of knowledge in the strict scientific sense, even though history proves these disparagements and accusations wrong. Despite criticisms, Sorokin continued to assert the need for the development of a system of knowledge appropriate to the times and suitable for a rapidly evolving society. In answering the criticisms advanced by Joseph B. Ford and Daya Krishna, he defended himself in *Reply to Criticisms of My Integral System of Knowledge* (Sorokin 1963b):

> "Such a system would include in it not only rational, sensory, and intuitive knowledge of rational-sensory realities but also the cognition of 'suprasensory and suprarational' forms of reality—the knowledge called 'no-knowledge' by the Taoist sages, *prajna* and *jnana* by the Hindu and the Buddhist thinkers, and *docta ignorantia* by Nicolas of Cusa. Development of such a genuine integral system of truth and cognition can greatly help mankind in enriching, deepening, and enlarging human knowledge of total reality, in eliminating the mutually conflicting claims of science, religion, philosophy, and ethics through reconciliation and unification of their real knowledge into one integral system of truth, in stimulating man's creativity in all fields of culture and social life, and in the ennoblement and transfiguration of man himself" (Sorokin 1963b: 400).

In *Fads and Foibles in Modern Sociology and Related Sciences* (1956), Sorokin had previously clarified some aspects of intuitive knowledge. The latter is an unforeseeable illumination of the researcher or the creative genius that provides him with the essence of the problem or its solution. It is essentially different from empirical-sensory knowledge and rational-mindful knowledge (logical-mathematical analysis and deductions). The supersensory and superational intuition is the exact opposite of the unconscious with which it is regularly confused. While the supersensory intuition is superior to rational (mental consciousness level), the subconscious or unconscious is below this level. While every cognition, discovery, or creative realization is always a knowingly automatic or superconscious fact, the unconscious, by definition, cannot discover or create anything other than instinct. Sorokin's claims, severely criticized by the majority of his contemporary scholars, were in fact backed up—in different words—by Charles Wright Mills. In the book *The Sociological Immagination* (1959), Wright Mills confirmed that we cannot understand the life of individuals without understanding society and vice versa (the reference here can be to the inseparability between personality, society, and culture). At the same time, he also argued that individuals need a quality of mind to help them use the information (which, given its amount, is increasingly difficult to assimilate) to develop a reason for achieving a lucid synthesis of what is happening or may happen to the individual and the world. This quality, called "sociological imagination", allows for a reading of biographies and history in mutual relation with society:

> "The sociological imagination enables its possessor to understand the larger historical scene in terms of its meaning for the inner life and external career of a variety of individuals. It enables him to take into account how individuals, in the welter of their daily experience, often became falsely conscious of their social positions. [...] The first fruit of this imagination—and the first lesson of the social science that embodies it—is the idea that the individual can understand his own experience and gauge his own fate only by locating himself within his period, that he can know his own chances in life only by becoming aware of those of all individuals in his circumstances" (Wright Mills 1959: 5).

In other words, intuitive knowledge or sociological imagination allows the scholar to move from one perspective to the other by seizing what is happening in the world. And at the same time to understand what happens to herself, and to individuals, as intersections of their biography and the history of society.

Reality is multifaceted and pluralistic, and it cannot ignore qualitative and quantitative aspects located in the three systems of knowledge (empirical-sensory, rational-mindful, and supersensory-superrational intuition). These systems are mutually connected by a causal-meaningful link. Their entirety provides the integral social reality (hence an integral system of sociology of knowledge, integralism or integral method) that includes the integral man. The integral being encloses in himself, in his existence, the superconscious creator, the rational thinker, the observer, and the empiricist. The truth obtained through the integral application of the three means of knowledge (sense, reason, and intuition) is a fuller and more valid reality than that achieved through only one of these means. In the integral method, these three means control and complement each other: "integral theory of truth cannot be reduced to any of the three forms of truth, but it embraces all of them" (Ford 1963: 52). The Integralism conjugates the truth of faith, the truth of reason, and intuition into an inclusive whole that allows for the creation of a framework for understanding life, the cosmos, and the role of humanity in it.

References

Barthes, R. (1964). *Le degré zéro de l'écriture, suivi de Eléments de sémiologie*. Paris: Ed. Gonthier.
Berger, P. L., & Luckmann, T. (1966). *The social construction of reality: A treatise in the sociology of knowledge*. New York: Penguin Books.
de Saussure, F. (1971). *Cours de linguistique générale*. Paris: Payot.
Ford, J. B. (1963). Sorokin as philosopher. In P. J. Allen (Ed.), *Pitirim A. Sorokin in review* (pp. 39–66). Durham: Duke University Press.
Homans, G. C. (1967). *The nature of social science*. New York: Hartcourt.
Johnston, B. V. (1999). Pitirim A. Sorokin on order, change and the reconstruction of society: An integral perspective. *Comparative Civilizations Review, 41*(3), 25–41.
Lévy, P. (1994). *L'intelligence collective. Pour une anthropologie du cyberspace*. Paris: Éditions La Découverte.
Luhmann, N. (1983). *Struttura della società e semantica*. Rome-Bari: Laterza.
Mangone, E. (2012). *Persona, conoscenza, società*. Milan: FrancoAngeli.
Mangone, E. (2015). *Knowledge for the future of Europe*. Cava de' Tirreni: Areablu edizioni.
Moscovici, S. (1984). The phenomenon of social representations. In R. M. Farr & S. Moscovici (Eds.), *Social representations* (pp. 3–69). Cambridge: Cambridge University Press.
Schütz, A. (1946). The well-informed citizen. An essay on the social distribution of knowledge. *Social Research, 14*(4), 463–478.
Sorokin, P. A. (1928). *Contemporary sociological theory*. New York: Harper and Brothers.
Sorokin, P. A. (1956). *Fads and foibles in modern sociology and related sciences*. Chicago: Henry Regnery.
Sorokin, P. A. (1957). *Social & cultural dynamics: A study of change in major systems of art, truth, ethics, law and social relationships*. Boston: Porter Sargent.

Sorokin, P. A. (1958). Integralism is my philosophy. In W. Burnett (Ed.), *This is my philosophy. Twenty of the world's outstanding thinkers reveal the deepest meaning they have found in life* (pp. 180–189). London: George Allen & Unwin.

Sorokin, P. A. (1962). *Society, culture, and personality. The structure and dynamics. A system of general sociology.* New York: Cooper Square.

Sorokin, P. A. (1963a). Sociology of my mental life. In P. J. Allen (Ed.), *Pitirim A. Sorokin in review* (pp. 3–36). Durham: Duke University Press.

Sorokin, P. A. (1963b). Reply to criticisms of my integral system of knowledge by professor Joseph B. Ford and Dr. Daya Krishna. In P. J. Allen (Ed.), *Pitirim A. Sorokin in review* (pp. 383–400). Durham: Duke University Press.

Wright Mills, C. (1959). *The sociological imagination.* New York: Oxford University Press.

Chapter 4
The Society and Its Paradoxes

1 Social Universe and Interactions

The transition from community to society[1] in the nineteenth and twentieth centuries was one of the most marked changes in a crisis situation. The latter generated a series of disruptions (including two world wars) that needed to find new forms and new settlements of sociocultural structures. Sorokin thus wrote, during World War II, in the book *The Crisis of Our Age* (1941): "the present crisis is not ordinary. It is not merely an economic or political maladjustment, but involves simultaneously almost the whole of western culture and society, in all their main sectors. It is a crisis in their art and science, philosophy and religion, law and morals, manners and mores; in the form of social, political, and economic organization, including the nature of the family and marriage—in brief, it is a crisis involving almost the whole way of life, thought, and conduct of Western society" (Sorokin 1941: 16–17). Today, despite the passing of almost a century, the form of societal disintegration that Sorokin described is still possible. In contemporary society, there are still changes in lifestyles, thinking, and acting that often do not attain new equilibriums.

Society, personality, and culture are the three elements that constitute the indivisible sociocultural trinity (Sorokin 1948), whose interactions determine the complexity of sociocultural phenomena. And it is precisely on social, cultural, and personal aspects that will focus attention in the next pages. We should point out, however, that considering them separately is a fabrication, since Sorokin explicitly declared them indivisible in *Society, Culture, and Personality*:

"The sociocultural order is indivisible, and no one can make a special science of one aspect of it, say, the social aspect, ignoring the cultural and personal aspects. [...] For this reasons 'society' cannot be a broader term than 'culture', nor can they be sharply separated from one another. The only possible differentiation is that the term 'social' denotes concentration

---

[1] This passage concerns the transformations of the foundations on which human aggregates are based: the community finds its foundation in relations of individuals, while society finds its foundation in the differentiation of the social roles.

© The Author(s) 2018
E. Mangone, *Social and Cultural Dynamics*, SpringerBriefs in Psychology,
https://doi.org/10.1007/978-3-319-68309-6_4

on the totality of interacting human beings and their relationships, whereas 'cultural' signifies concentrations on meanings, values, and norms and their material vehicles (or material culture) (Sorokin 1962: 65).

And yet their "individual" discussion is necessary in order to be able to better understand their peculiarities, so as to actualize them to current times. This analysis begins with society understood as the whole of the personalities interacting with their social and cultural processes.

"What is commonly called society, or social life, actually consists of an immensely complicated that its study and comprehension as a whole is a sheer impossibility. It would be futile, indeed, to attempt to grasp in entirety the infinitely heterogeneous mass of social events, facts, acts and relations without some preliminary simplification. The student of society will do well, then to begin his research by breaking it up into its constituent parts and directing his attention first to the most generic phenomenon, in which the social expresses itself. If such a phenomenon, small as it may be, will be found to exhibit the fundamental characteristics of social life, it can then be used as a model for the study of the more complex social phenomena as the primary social fact—the fundamental unit of sociology. [...] What particular phenomenon in the realm of human relations will serve then as this simple unit, as a simplified model of social life in all its vastness and complexity? [...] no single individual can constitute a 'society' nor can a 'social phenomenon' [...] A social phenomenon demands the existence of a number of individuals—at least 2. However, mere independent existence will not suffice. In order for individuals to constitute a 'society', create a 'social phenomenon', give rise to a 'social process', it is necessary that they interact—exchange actions and reactions (Thus, Crusoe and Friday formed a 'society'; their interrelations were 'social interactions'; their actions and reactions, 'social processes'). Only then can their life be called social (Sorokin n.d.: Chap. IV, pp. 1–2).

In other words, the focus is on individuals. Without individuals, society cannot exist, as it is based on their interactions that generate social meanings and processes. Sorokin himself, in the undated document *The Nature of Sociology and its Relationship to Other Sciences*, when describing the social universe, spends a long time detailing the main forms of sociocultural interactions. Sociocultural interactions can be classified into a few classes based on some key aspects: (1) the number and properties of interacting individuals; (2) the character of actions by individuals; and finally, (3) the nature of the transmission vehicles. Interactions occur between individuals (interpersonal interaction) and between organized groups (inter-group interaction). So, based on the number of individuals involved, interactions take on different shapes ranging from those occurring between two individuals, between many individuals, from one to many, from many to many. The same happens for groups. In the book *Society, Culture, and Personality* (1962), Sorokin describes in detail the properties, the characters, the forms of interactions, and the characteristics of the conductors through which the interactions are carried.

Regarding the properties of the interactions, Sorokin states that there is ample variety of them, but since they cannot be all examined, he considers only "one characteristic, namely, the biopsychological and sociocultural *homogeneity or heterogeneity* (similarity or dissimilarity) of the interacting individual or groups" (Sorokin 1962: 43). With respect to the character of the individual actions that respond to an external stimulus or another, the process of interaction can be classified in thousands of ways. For mere knowledge purposes, we list the four main forms identified by

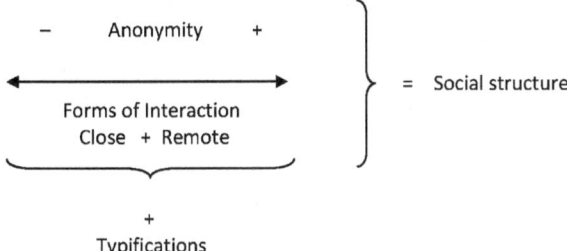

Fig. 4.1 Determination of the social structure

Sorokin: those that exert an influence merely through the *known existence* of the party or parties (the *catalytic* form); through performing *overt actions*; through *abstention from overt action*; through active *toleration*. Instead, the conductors of interactions (vehicles) can be distinguished into: "Physical conductors are those in which the physical qualities of the vehicle are used to modify the state of mind and the overt actions of another. [...] Symbolic conductors, exert an influence not so much by virtue of their physical properties as by virtue of the symbolic meaning attached to them" (Sorokin 1962: 52–53). In summary, society cannot exist "beyond" individuals and "independently" from them, but only as a system of interacting units in the "world of everyday life". The latter becomes the "place", metaphorically speaking, in which to analyse the interactions and, above all, the social reality. Within it, the intentionality towards objects directs the actions of individuals. These objects, in turn, through differentiation, appear to consciousness as constituents of different "spheres of reality".

Let us now call up a few notions by Berger and Luckmann (1966). When interaction occurs through a face to face encounter, we are dealing with an interaction that can be called "*close*" because it allows a personal and direct relationship (encounter of subjectivity). Conversely, when interactions occur in other forms, they do not allow the encounter with the subjectivity of the other ("*remote*"). All forms of interaction therefore influence one another, starting from the meanings we attribute to them. These are based on typologies that undergo a continuous negotiation process based on the more or less direct interactions. In fact, interactions become the more *anonymous* the more they move away from situations, but the degree of anonymity is also influenced by the intimacy of interaction.

The sum total of the forms of interaction and of the typings appearing in recurring patterns determines the social structure of society (Fig. 4.1), which is an essential element of the reality of everyday life. Interactions with others are temporally located: they are not limited to contemporaries, but may also refer to predecessors or successors (present, past, and future). It follows that human beings do not interact only with their natural environment, but also and above all with the sociocultural one in which they act and from which they are subject to a constant socially determined influence. The development of everyday life is therefore explicit in a series of "*routines*" that reduce the moments of analysis and redefinition of a situation. We never start from scratch to make a decision. Routines allow us to save precious energy in the event of a situation requiring a "deliberative" and "innovative" decision-making. These processes always precede a process of *institutionalization*: "In actual experience institutions generally manifest themselves in collectivities containing considerable numbers of people. It is theoretically important, however, to emphasize that the institutionalizing process of reciprocal typification would occur even if two individuals began to interact de novo. Institutionalization is incipient in every social situation continuing in time" (Berger and Luckmann 1966: 73). The institution is external to individuals, and yet it is produced by them through a social dialectic consisting of three moments (externalization, objectivation, and internalization). Individuals are not born as members of society, but they are born with a predisposition to sociality that allows them to become a member of society through subsequent stages in temporal succession. "The beginning point of this process is internalization: the immediate apprehension or interpretation of an objective event as expressing meaning, that is, as a manifestation of another's subjective processes which thereby becomes subjectively meaningful to myself. This does not mean that I understand the other adequately" (Berger and Luckmann 1966: 149). The externalization consists of two determined and successive time arcs. It is the instance when individuals first form their basic knowledge and define their own expectations (first time span), then recreate attitudes and lifestyles by virtue of their knowledge (second time span). The objectivation allows individuals to perceive the consequences of their actions towards themselves and others. To further clarify this "dialectical process" between individuals and social reality, we can say that reality does not affect individuals for what it is but for what they consider it to be. Society must thus be included in this dialectical process, which is composed of these three moments. These are to be understood as phases of a chronological sequence that for individuals begins with the internalization realized through the socialization process (Dubar 2003). The objective reality of the institutions, being a construction of human beings, requires a legitimacy that does not tell individuals only what they should do but also explains why things are as they are. In this way, knowledge precedes values in the legitimization of the institutions as previously defined. The legitimization process takes place at different levels. The highest level (the fourth) is that of the symbolic universes resulting from the signification processes. In other words, the symbolic universes hold together what Sorokin identifies as the three components of sociocultural phenomena and the processes of meaningful human interaction: the human beings interacting, the meanings (cognitive, values, and norms), and the vehicles (psychics and symbolics).

These processes are related to other realities than those of everyday experience and they succeed in combining the function of legitimizing both the biographies of individuals and the institutions (they build a hierarchy of reality within everyday life). The symbolic universe is the matrix of all socially objective and subjectively real systems of meaning. The existence of the entire society in the historical sense and that of the individuals fully occur within these universes. Through the process of legitimization, they allow the incorporation of the experiences of different spheres of reality into the same symbolic universe. Given their ability to integrate the institutional (public) sphere and individuals (private sphere), symbolic universes are a "protective dome" that covers the institutional order as well as the biographies of individuals. They circumscribe social reality and set what is relevant to social interaction.

Sorokin first, and Berger and Luckmann later—with very different arguments, but similar in some respects—reach the same conclusion: what is commonly called society or social life actually consists of a complex heterogeneous set of social events, facts, acts, and relationships between them. For this reason, the analysis of the properties of sociocultural phenomena should not be considered as the "simplest unit".

2 The Construction of Reality and the Interactions (Social Relations)

The multidimensionality and multiplicity of factors influencing the actions of individuals prompt us to say that the concept of reality shows a high degree of subjectivity. This is especially related to some aspects of the culture and the indissoluble link with the biography of the individual. Reality should be understood not as "constitutive", but as a "social fact" deriving from the historical forms through by which occur the communication between social system and lifeworld.

Reality is not stable, nor is it common to all individuals: everyone represents the "world" in a different way. This diversity develops on the basis of the interactions experienced in everyday life and of socially derived knowledge[2] (Schütz 1946) as well as of relational reflexivity (Donati 2011). Thus, changes in individuals' actions are the result of the combination of the three dimensions of body, mind, and soul (Sorokin 1963), and of the wider understanding of the social environment within which they act. As Sorokin already stated (n.d., Chaps. IV–VI), relationships take on a variety of shapes in the complex scenario of society. The social universe (society) is therefore the sum of the social lives of individuals. In turn, the latter are the result of the reciprocal conditioning of individuals in interaction (social relationship) and are based on relative autonomy that also determine the close connection between the three elements of the indivisible sociocultural trinity (Sorokin 1948, 1962)—personality, society, and culture. Maintaining Sorokin's theoretical framework—the triangulation of individual, social and cultural dimensions—we outline the forms of the relationships between individuals in relation to the dynamics of everyday life occurring in contemporary society. There are four relational forms that bud and develop along a subjective dimension and an institutional one. They are characterized by a common aspect: the man–environment pair (historically differentiated) has seen the prevalence of one or the other term, but the disposition has always been oriented towards the survival and development of the human species (with positive and/or negative effects on the community).

---

[2] Only a small part of an individual's knowledge is actually the fruit of his experience; most of it derives from others who have already had that experience and handed it down.

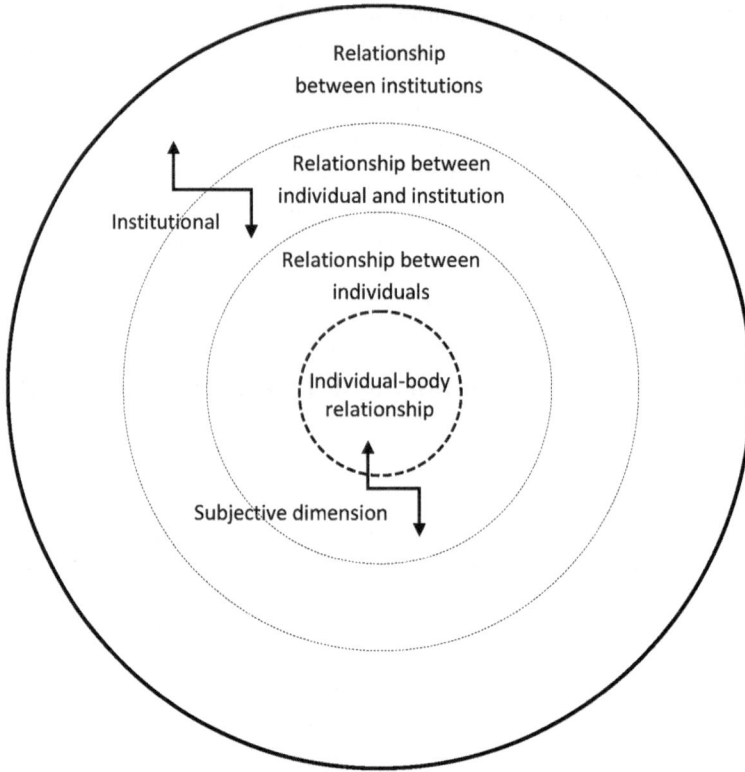

Fig. 4.2 Representation of the relationships linked to the institutional and subjective dimensions

The proposed forms of expression can be imagined as a sequence of concentric circles[3] (Fig. 4.2) moving from the inside to the outside along the subjective–institutional axis. They are: (a) individual–body relationship; (b) relationship between individuals; (c) relationship between individual(s) and institution(s); (d) relationship between institutions. The image takes shape from the innermost circle (individual–body relationship) from which all other forms of relationships branch out. Broadly following Boudon's *Weberian paradigm*[4] (1984), we can state that all these circles are generated seamlessly from the centre. Their subdivision is artificial, aimed at simplifying their description and understanding, as readers can benefit from a synthetical graphical representation.

The innermost circles refer to the relationship between individuals and to that between individual and body. It should be noted that the body should not be understood as *Körper* (organic body), but as *Leib*, that is, a living body in the world

---

[3] We proposed a variation of this figure in the article *Social relationships and health* (Mangone 2013).

[4] Boudon's Weberian paradigm states that a specific phenomenon derives from the actions depending on the situation that generated them, which in turn depends on other phenomena.

(Husserl 1970). The subject experiences his own body because the individual-body relationship brings with it a holistic perspective in which daily life is expressed and is also lived through the body (Hepworth and Turner 1991). This form of interaction allows each individual to build her biography, her life story through the body, which is the "principal instrument" through which an individual perceives what is "external" and "alien". We should remember that Sorokin, when talking about human beings, always referred to them as composed of three elements including the body (body, mind, and soul). The body is a *structure* that binds one to what is "outside". It is the *means* by which the subject can communicate (including communicate oneself), and know (one of the three truth systems found by Sorokin is based on the sensations that are mostly acquired through the body). In contemporary society, which tends towards uniform globalization, the body becomes the instrument of expression of the living world because it is not mediated, communicating the state of the individual beyond the spoken language (Watzlawick et al. 1967). Another important aspect is that from the inevitable individual-body relationship comes also the discovery of the relationship with the Other, with other individuals. The constant pursuit of a balance between internal and external forces an individual to relate to the world around her, and this, intentionally or not, means dealing with other individuals (the second circle).

In an interaction context, two important aspects cannot be ignored: the positions and roles of the individuals involved and the intensity of the relationship (Cooley 1962). In this way, the attention is shifted to the institutional axis that focuses on the relationships falling into the secondary type (prevalence of the role), while the forms of the relationships that lie on the subjective dimension fall into the primary type (subjectivity).

The secondary interaction that can be established between an individual and the institution causes a change of frame (Goffman 1981). In this case the experience occurs within the institutional system, where the role of the individual prevails over that of the subjectivity. To make the arguments more understandable, we could say that the institution has a role responsibility and a specific function. Individuals who come into contact with an institution are considered, mostly, as "anonymous" units addressing a system working in a technically competent way. The institutional dimension of the relationship emerges overwhelmingly when it comes to interaction between institutions. Here the preponderance of the role is absolute. This is especially true for institutions with a strong hierarchy that implies a decrease in participation in decision-making for lower levels.

In light of the above, and keeping in mind that society is "the totality of interacting personalities, with their sociocultural relationships and processes" (Sorokin 1962: 63), interactions are the process that relate individuals to their life world. In this framework, institutions should be open systems. They should propose a model of interaction essentially based on the opening between internal and external. This would allow to take due account of Sorokin's indivisible sociocultural trinity (1948)—individual, society, and culture—and of the environment of reference (both physical and social).

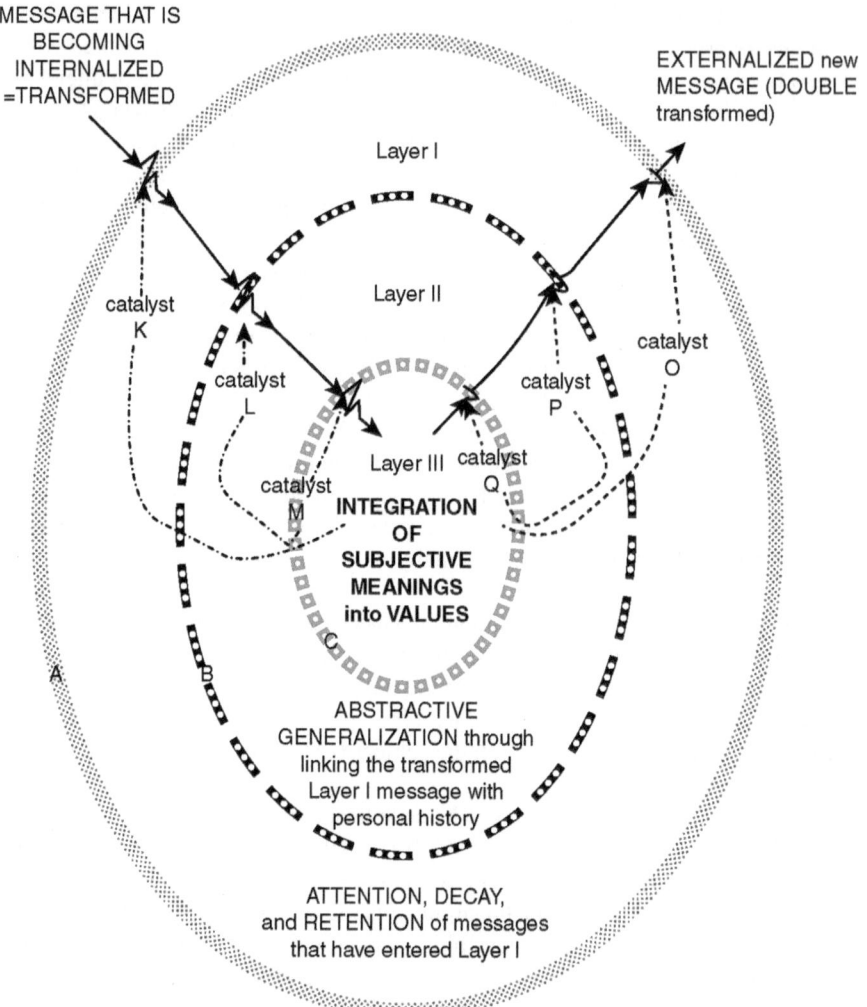

MESSAGE THAT IS
BECOMING
INTERNALIZED
=TRANSFORMED

EXTERNALIZED new
MESSAGE (DOUBLE
transformed)

Layer I

catalyst
K

Layer II

catalyst
O

catalyst
L

catalyst
P

Layer III   catalyst
Q

catalyst
M

**INTEGRATION
OF
SUBJECTIVE
MEANINGS
into VALUES**

C

A   B

ABSTRACTIVE
GENERALIZATION through
linking the transformed
Layer I message with
personal history

ATTENTION, DECAY,
and RETENTION of messages
that have entered Layer I

Fig. 4.3 Laminal Model of internalization/externalization (Valsiner 2014: 71)

The above discussion has obviously been approached through a sociological interpretation. But if we think of Sorokin's integrated system of knowledge, Integralism
(1958), we have to consider other perspectives, so as to have a vision of the whole as
broad as possible. We thus forward the perspective of cultural psychology (2014). In
particular, we refer to the laminal model (Valsiner 1997) of the internalization/exter-
nalization process (Fig. 4.3) whose sequence of boundaries demarcates personal
(internal) aspects and aspects related to the outside world. There are many similarities. For example, if we compare this model to the previous one, the interior and
exterior may correspond to the subjective and institutional dimension. In addition,
both models have (labile) boundaries separating the various dimensions.

If the above is a key to read individual interactions, from which processes and meanings arise, we cannot deny that human knowledge interacts with social life. Ideas on what is right and what is not are vastly predetermined: the cultural and social environment within which they develop create "cultural mentalities" as defined by Sorokin (1957), regardless of what is commonly accepted or rejected.

The social universe appears as a boundless network of interactions that do not always have a linear logic. Rather, they are often ambiguous, oscillating between information exchange and symbolic action on other individuals. The forms of interaction that subjects implement and experience are a problematic act that often does not leave room for individuals' equality and/or reciprocity.

3 The Paradoxes of Contemporary Society

The history of humanity has always characterized the orientation and study approaches of social sciences. These were distinguished by the contrasts individual societies: yesterday/today, normal/pathological, north/south, friend/enemy, centre/periphery, rich/poor, and so on in a list that could go to infinity. The new millennium, however, in the wake of the last decades of the twentieth century, seems to have removed all contrasts. They have not actually been eliminated, but the idea of society has been reformulated. Today, we talk of a "world society".

Wallerstein (1976), more than 30 years ago, coined the concept of "world system". He argued that the ability of mankind to intelligently participate in the evolution of its own system depends on its ability to understand the latter's totality—recalling Sorokin's idea of the indivisible sociocultural trinity (1948). For social sciences, this means focusing on the processes of structuring and destructuring, integration and exchange, external conflict and internal reproduction of the economy, politics, culture, and the community system. These are seen and read as subsystems of a society that seems to have no physical boundaries anymore, and recognizing oneself beyond boundaries involves dealing with a complex interaction process. On the one hand, this states—on an informal level—the existence of a community that can be actively involved in new opportunities for movement and reciprocity; and, on the other, it reaffirms the conditions resulting from the legitimate division of a territory. In this developmental logic, however, there are some paradoxical aspects that scholars often overlook and that Sorokin partially highlighted in the last century in *The Crisis of Our Age* (1941). To clarify, at least in part, some of these aspects, we describe the two paradoxes most affecting the elements of contemporary society: the first relates to the local appropriation of information in the face of its global diffusion, and the second concerns the equity principle.

Information dissemination processes occur on a global scale, but the appropriation of information takes place locally because individuals' lives happen in a precise and definite space and time that produce sociocultural causalities (Sorokin 1964). The appropriation of media-generated information, for example, is always a localized phenomenon. It involves individuals who interpret and incorporate information into their lives using the resources they find in the contexts they live in. The global-ization of communication has not erased the local character of appropriation, instead it has created a new kind of symbolic universe of the modern world (Thompson 1995). This symbolic axis—global diffusion vs. local appropriation—is character-ized by the acquisition of information, images, knowledge, and other artefacts with the typical modes and ways of globalized society. However, the latter are interpreted and elaborated in the places where individuals lead their daily lives, usually aiming at consolidating values and beliefs. All this focuses the attention to the transforma-tion of the conceptual structure of space-time distancing. Giddens (1990) argued that in modern times this spacing is much larger than at any other time, since social forms and events, both local and distant, are "stretched".[5] If this is how millions of individuals live today, we can think it a paradox. In the face of a global dissemina-tion of information, knowledge, images, and symbols that invest all fields of social life (e.g. from economics to communication), there is a local adaptation and elabo-ration enhancing differences and cultural divisions. Such a context produces a fur-ther perverse effect (Boudon 1977). The defence of the local dimension increases also through ethnocentric actions and attitudes, which aim mainly at reappropriat-ing the individual and collective identity uniformized by excessive mass media information. The same information that contribute significantly to the construction of social reality and to the establishment and definition of the self as a path and a project of individual life. The end result of these processes is a greater fragmenta-tion of the world (Geertz 1996). In fact, in the third millennium, despite the unheard-of progresses in knowledge and technology—especially in the Western world—there are doubts about an uneven development and diffusion of knowledge. This entails an increase in inequalities and the consolidation of subordination relationships by some peoples over others. Such doubts seem today to be certainties.

These changes seem to accentuate the cultural (sensate) mentality, the same to which Sorokin attributed the disintegration of solidarity links in favour of an exas-perated individualism that would led to the destruction of humanity and not to its possible escape from the crisis, as he suggested in *The Reconstruction of Humanity* (1948).

Consequently, the change in contexts and their greater complexity has called for a reflection on the issue of inequalities. This prompts the reflection on another paradox of contemporary society: equity.[6] Equity in itself can be understood in dif-ferent forms, as various and different are the methods for applying this principle. It can be understood as (1) an equitable distribution of resources between different groups (social ones, ethnic ones, etc.); (2) the equal access to resources regardless of the person's income; and finally, (3) an equal access opportunity for equal needs.

---

[5] Giddens emphasized the complex relationships existing between circumstances of coexistence (local implications) and connections of presence and absence (distance interaction).

[6] Beyond potential semantic differences in different languages, in order to understand the socio-economic and political dynamics of contemporary society it is sufficient to attribute to the term "equity" the generic meaning of a final distribution of resources more equal than to that originally derived from economic and financial markets.

No model of welfare system currently in force throughout the world (Esping-Andersen et al. 2002) has succeeded in securing and guaranteeing the principle of equity by combining these different forms. It is evident for everyone that this political project is still far from being real and maybe it will never be. Equity is not saying "everything for everyone" but "what is needed for everyone to have equal chances of choosing for their own life project". In other words, we are back to Sen's *functioning* and *capabilities* concepts (1982, 1987). Referring Dahrendorf's arguments (1988), we can say that *life chances*—understood as a choice between alternatives—are never distributed equally. There are no societies in which all individuals have the same *entitlements* (access and legitimate control over things) and enjoy the same *provisions* (set of material and immaterial choices). The fundamental problem is therefore the distribution and evolution of the choices that must be the product of mediation that leads to a high level of compatibility. Choices can be taken freely only if opportunities are equally distributed. Distribution should be guided by individual needs rather than by social privilege. This methodology is very close to the Aristotelian concept of "vertical equity" mentioned in *Ethica Nicomachea*. In this case different individuals must be treated differently, while similar individuals must be treated equally ("horizontal equity").

To date, in a society that has globalized all processes, no government is able to apply the principle of equity in the Aristotelian sense. The future challenge is combining the scarcity of resources with the interests of improving the living conditions of individuals. For Sorokin, the future of mankind and its development is in the hands of mankind itself, neither the law nor the education, neither religion nor the economy, or the science can be sufficient for this task. This arduous task is for the humanity. "A peaceful, harmonious, and creative society can exist only when its members possess at least a minimum of love, sympathy, and compassion ensuring mutual aid, co-operation, and fair treatment. Under these conditions its members are united in one collective 'we' in which the joys and sorrows of one member are shared by others" (Sorokin 1948: 57). All this cannot ignore the need to address the theme of social interactions (individual and collective) that represent the essence of existence.

References

Berger, P. L., & Luckmann, T. (1966). *The social construction of reality: A treatise in the sociology of knowledge*. New York: Penguin Books.

Boudon, R. (1977). *Effects pervers et ordre social*. Paris: PUF.

Boudon, R. (1984). *La place du désordre. Critique des théories du changement*. Paris: PUF.

Cooley, C. H. (1962). *Social organization: A study of the larger mind*. New York: Schocken Books.

Dahrendorf, R. (1988). *The modern social conflict. An essay on the politics of liberty*. New York: Weidenfeld & Nicolson.

Donati, P. (2011). Modernization and relational reflexivity. *International Review of Sociology, 21*(1), 21–39.

Dubar, C. (2003). *La socialisasion. Construction des identités socailes et professionalles*. Paris: Armand Colin.

Esping-Andersen, G., Gallie, D., Hemerijck, A., & Myles, J. (Eds.). (2002). *Why we need a new welfare state*. Oxford: Oxford University Press.

Geertz, C. (1996). *Global world, local worlds*. New York: Basic Books.

Giddens, A. (1990). *The consequences of modernity*. Cambridge: Polity Press.

Goffman, E. (1981). *Forms of talk*. Philadelphia: University of Pennsylvania Press.

Hepworth, M., & Turner, B. S. (Eds.). (1991). *The body. Social process and cultural theory*. London: Sage.

Husserl, E. (1970). *The crisis of European sciences and transcendental phenomenology: An introduction to phenomenological philosophy*. Evanston: Northwestern University Press.

Mangone, E. (2013). Social relationships and health. *Salute & Società, 12*(3), 193–206.

Schütz, A. (1946). The well-informed citizen. An essay on the social distribution of knowledge. *Social Research, 14*(4), 463–478.

Sen, A. (1982). *Choice, welfare and measurement*. Oxford: Blackwell.

Sen, A. (1987). *Commodities and capabilities*. Oxford: Oxford University Press.

Sorokin, P.A. (n.d.). *The nature of sociology and its relation to other sciences*. University of Saskatchewan: University Archives & Special Collections, P.A. Sorokin fonds, MG449, I, A, 3.

Sorokin, P. A. (1941). *The crisis of our age. The social and cultural outlook*. New York: E.P. Dutton.

Sorokin, P. A. (1948). *The reconstruction of humanity*. Boston: The Bacon.

Sorokin, P. A. (1957). Social & cultural dynamics. In *A study of change in major systems of art, truth, ethics, law and social relationships*. Boston: Porter Sargent.

Sorokin, P. A. (1958). Integralism is my philosophy. In W. Burnett (Ed.), *This is my philosophy. Twenty of the world's outstanding thinkers reveal the deepest meaning they have found in life* (pp. 180–189). London: George Allen & Unwin.

Sorokin, P. A. (1962). *Society, culture, and personality: Their structure and dynamics. A system of general sociology*. New York: Cooper Square.

Sorokin, P. A. (1963). Sociology of my mental life. In P. J. Allen (Ed.), *Pitirim A. Sorokin in review* (pp. 3–36). Durham: Duke University Press.

Sorokin, P. A. (1964). *Sociocultural causality, space, time. A study of referential principles of sociology and social science*. New York: Russell & Russell.

Thompson, J. B. (1995). *The media and modernity. A social theory of the media*. Cambridge: Polity Press.

Valsiner, J. (1997). *Culture and the development of children's action*. New York: Wiley.

Valsiner, J. (2014). *An invitation to cultural psychology*. London: Sage.

Wallerstein, I. (1976). A world-system perspective on the social science. *The British Journal of Sociology, 27*(3), 343–352.

Watzlawick, P., Beavin, J. H., & Jackson, D. D. (1967). *Pragmatic of human communication. A study of interactional patterns, pathologies, and paradoxes*. New York: W.W. Norton.

Chapter 5
The Cultural System and the Social Problems

1 The Cultural Universe

Culture is not a static element; indeed, it is constructed and reconstructed based on an endless definition process. This process gives us the opportunity to identify the shared values and attitudes supporting both structure and actions—what Sorokin called "cultural mentality" (1957). After all, cannot forget that the ways in which human action is manifested are culturally determined and also filtered through the approval of group. In this case culture encompasses the tools (language, symbols, signs, etc.) that provide meaning as they are shared within a context that must then validate the action.

It is not by chance that cultural systems are an element of the indivisible socio-cultural trinity (Sorokin 1948), together with society and personality. Sorokin (1962) identifies them as indispensable and indivisible elements in the analysis of sociocultural phenomena. However, the author does not provide a precise definition of what "human culture" means. He thinks it unnecessary to give a restrictive definition, instead preferring a generic one: *"In the broadest sense it may mean the sum total of everything which is created or modified by the conscious or unconscious activity of two or more individuals interacting with one another or conditioning one another's behaviour.* According to this definition, not only science, philosophy, religion, art, technics, and all the physical paraphernalia of an advanced civilization are cultural phenomena" (Sorokin 1957: 2).

However, culture has precise features that cannot be ignored in an integral system of knowledge rationale. Cuche (1996) offers a definition based on these features, stating that the expression of the totality of man's social life. It is characterized by its collective dimension. In the end, culture is acquired and therefore does not depend on biological heredity. However, although culture is acquired, its origin and its characteristics are predominantly unconscious. In other words, culture is constituted by both objective elements (tools, capacities, etc.) and subjective elements (beliefs, roles, values, etc.). It is one of the main factors for assessing people's

© The Author(s) 2018
E. Mangone, *Social and Cultural Dynamics*, SpringerBriefs in Psychology,
https://doi.org/10.1007/978-3-319-68309-6_5

compliance (integration) with society. All activities and institutions are "cultural", since their functioning requires meanings to be made explicit. It follows our disagreement with the idea that societal living is linked to cultural determinism. Rather, culture is seen as a central component for individual actions: "every social practice depends on and relates to meaning; consequently, that culture is one of the constitutive conditions of existence of that practice, that every social practice has a cultural dimension" (Hall 1997: 225–226). Cultural objects hold significance among the people who live within a given social world and the latter, in turn, has meaning only through the culture (Griswold 1994) with which it is observed. Therefore, "culture is a fundamental aspect of daily life and, as such, it is necessary to understand it in relation to the different situations of the social world. Through this study, possible pathways to improve relations can be hypothesized and the forms deriving from this social world, and through which interactions between people and the other elements of the system are expressed, can be improved" (Mangone 2015: 45).

From these definitions, however, it follows that not everything is a "cultural phenomenon". This becomes clearer if we bear in mind that Sorokin considers social sciences as dealing with superorganic phenomena, i.e. those phenomena that are found only in man and the man-made world. Indeed, "The cultural aspect of the superorganic universe consists of meanings, values, norms, their interaction and relationships, their integrated and unintegrated groups ("systems" and "congeries") as they are objectified through overt actions and other vehicles in the empirical sociocultural universe" (Sorokin 1962: 313). In this way, individuals (or groups) play the main function of both agents and instruments of meanings, values, and norms, i.e. of producers of meaningful interactions. Culture is a set of meanings, values, and norms that are mutually correlated even when they seem differentiated (subcultures).

> "The cultural aspect of meaningful interaction consists of (1) the totality of meanings, values, and norms possessed by the interacting individual and groups, making up their "ideological" culture; (2) the totality of their meaningful actins-reactions through which the pure meanings, norms, and values are objectified, conveyed, and socialized, making up their behavioural culture; (3) the totality of all the other vehicles, the material, biophysical things and energies through which their ideological culture is manifested, externalized, socialized, and solidified, making up their "material" culture. *Thus the total empirical culture of a person or group is made up of these three levels of culture: ideological, behavioural, and material*" (Sorokin 1962: 313).

Ideological cultures are understood as "the integrated systems of science, philosophy, religion, ethics, law, the fine arts and the system of oral and written language as the main vehicle for objectification of any system or congeries of meanings" (Sorokin 1962: 317). The relationship between how individuals think and perceive reality, one the one hand, and society's system of thought, on the other, is a complex interaction process. In it, language is a form of objectivation of human expression. The term "dialectical process" is not without reason, and dialectics presuppose the presence and use of a language. The latter is the means through which the intellect can know the outside world and understand the experience of the human spirit. With the evolution of the species (*Homo sapiens*) we enters what is considered to be the age of the word, that will be later recognized as the age of language (Ong 1982; Goody 1986). Language should be understood as a form of universal symbolic mediation through which the different domains of meaning come to exist. Language, therefore, on the one hand, is what allows to modify opinions; and on the other, it becomes how we show our harmony with—or distance from—others. However, we must point out that it is necessary to consider separately those cultures based solely on orality (primary cultures)—based on what is remembered (memory)—which is the only feasible way for an individual to experience knowledge—from those in which writing has been established, changing communication and social organization to the core. Considering Thompson's theorization (1995) and the characteristics of orality, it is possible to associate the latter with the "face-to-face interaction" that requires coexistence in space and time. Conversely, writing can be associated with the "mediated interaction" as it requires the use of a technical medium (and in the specific case of paper and pen, or of a computer).

Addressing the three levels of culture that Sorokin proposes in *Society, Culture and Personality*, he states that, "when the integrated ideology of the group is consistently practiced in the overt behaviour of its members. The culture of the group becomes triply integrated when the integrated ideology of the group is adequately realized in the behaviour and vehicles of the group. In that case the integration of group's culture becomes complete, meaningful and causal, integrated in its ideological, behavioural, and material culture, with the members solidary in their ideology, behaviour, and vehicles" (Sorokin 1962: 323). It is by now also clear that the quarrels between those proclaiming the supremacy of the individual on society and those who, vice versa, affirm the supremacy of society on the individual is futile. The individual, social, and cultural components are interdependent and mutually influencing. Therefore,

"the mutual correspondence of the three aspects of the superorganic: (1) A given individual manifests his social and cultural universe. (2) A culture mirrors its human members and their group organizations. (3) A social structure reflects its component individuals and their cultural patterns. [...] The categories of the *individual*; the *social* and the *cultural* are indeed three inseparable aspects of the same superorganic phenomenon. This inseparableness is well manifested by (a) the concomitant development of the individual mind and the social structure; (b) close correspondence between the structure of the individual's egos and the structure of the groups to which the individual belongs; (c) by the effective determination of the individual's conduct by the groups in which he lives; (d) by the general defining of the content of the individual's mind (his scientific, philosophical religious, ethical, juridical, aesthetic and other values) by the cultural world in which he moves" (Sorokin 1962: 343).

As previously stated, all that is related to the cultural dimension is dynamic rather than static. And the first characteristic of this dynamism is how these systems of meanings, norms, and values are born, reproduced, and embedded by individuals and groups. However, some emerging elements are common to all cultural systems: "Any empirical cultural systems or congeries passes through three fundamental phases: (1) the conception (invention, creation, and unification) of two or more meanings, values, and norms to form a consistent system or congeries; (2) the objectification of the ideological system or congeries in the vehicles; (3) its socialization among human beings in either an ideological only or behavioural and material forms" (Sorokin 1962: 537). These three moments are quite similar to those introduced some years later by Berger and Luckmann (1966) on the institutionalization process: externalization, objectivation, and internalization.

Human beings, in their daily life experience, are producers of meanings, norms, and values, because they have meaningful interactions. We remind the reader, however, of the cyclical movements the cyclical movements of the systems (Ideational, Sensate, and Idealistic) identified by Sorokin (1957), that are produced by the transformations of the mental bases of individuals and characterized by particular values and forms of knowledge. The cultural mentality concerns, the experience related to individual thinking and the symbolic mediation processes allowing for the attribution of meaning. Since the attribution of meaning is an ongoing, evolving process, there are endless changes and transformations. Therefore, all cognitive activities (construction of conceptual and representative maps) can be considered as reaching a balance between the *assimilation* process that ensures continuity over time (understood as *Chrónos*—past, present and future) and the *accommodation*[1] process, understood as the ability to deal with future changes and uncertainties.

2   Mentality of Culture (or Culture Mentality)

The above-described aspects highlight the fundamental importance of the two dimensions of time and space in characterizing the occurrences of everyday life. And it is along the temporal and spatial axis that Sorokin describes, in detail and with innumerable examples (almost to the point of boring the reader), the integrated cultural systems deriving from the cultural mentalities that have evolved over the course of history (theory of cyclical movements).

To see the different forms that cultural mentalities have taken over the centuries, we use Sorokin's description:

"The elements of thought and meaning which lie at the base of any logically integrated system of culture may be considered under two aspects: the *internal* and the *external*. The first belongs to the realm of inner experience, either in its unorganized form of unintegrated images, ideas, volitions, feelings, and emotions; or in its organized form of systems of thought woven out of these elements of the inner experience. This is the realm of mind, value, meaning. For the sake of brevity we shall refer to it by the term "mentality of culture" (or "culture mentality"). The second is composed of inorganic and organic phenomena: objects, events, and processes, which incarnate, or incorporate, or realize, or externalize, the internal experience. [...] This means that for the investigator of an integrated system of culture the internal aspect is paramount. It determines which of the externally existing phenomena—and in what sense and to what extent—become a part of the system. In other words, it controls the external aspect of the culture. Deprived of its inner meaning, the

---

[1] The term *accommodation* refers to the transformations undergone by past experiences when a person receives new information, while the term *assimilation* indicates the degree to which any information received from the environment may be more or less adapted to the background of experience already possessed by the person.

Venus of Milo becomes a mere piece of marble identical in its physicochemical qualities with the same variety of marble in the state of nature. A Beethoven symphony turns into a mere combination of sounds, or even into a vibration of air waves of certain lengths to be studied by the laws of physics" (Sorokin 1957: 20–21).

The theory of cyclical movements of mental systems (Ideational, Sensate, and Idealistic) derives from the transformations of the mental bases of men and groups. These transformations give rise to multiple systems of logically integrated cultures. Each of them has its own mentality, its own system of truth and knowledge, its own philosophy, and a *Weltanschauung*. In other words, they have their own characteristics, but since culture is characterized by its inner aspects (mentality), to study these systems, one has to start with their elements. To study culture mentalities, we need to examine the following four items: "(1) *the nature of reality;* (2) *the nature of the needs and ends to be satisfied;* (3) *the extent to which these needs and ends are to be satisfied;* (4) *the methods of satisfaction*" (Sorokin 1957: 25). The definitions and characteristics of the three systems identified by Sorokin (Ideational, Sensate, and Idealistic) will be here summarized, referring curious readers to *Social & Cultural Dynamics* (1957)[2] for the meticulous descriptions and the historical and geographical examples related to the various areas of everyday life,[3] as well as theoretical and methodological problems.

All ideational culture systems—in relation to the four items mentioned above—are characterized by the following common premises: (1) reality is perceived as nonsensate and nonmaterial, everlasting Being *(Sein)*; (2) the needs and ends are mainly spiritual; (3) the extent of their satisfaction is the largest, and the level, highest; (4) the method of their fulfilment or realization is self-imposed minimization or elimination of most of the physical needs, and to the greatest possible extent (Sorokin 1957: 27). Despite these common premises, if we take into account only the variations in item (4), ideational systems can be further distinguished in:

"**A.** *Ascetic Ideationalism.* This seeks the consummation of the needs and ends through an excessive elimination and minimization of the carnal needs, supplemented by a complete detachment from the sensate world and even from oneself, viewing both as mere illusion, nonreal, nonexisting. The whole sensate milieu, and even the individual "self," are dissolved in the supersensate, ultimate reality. **B.** *Active Ideationalism.* Identical with general Ideationalism in its major premises, it seeks the realization of the needs and ends, not only through minimization of the carnal needs of individuals, but also through the transformation of the sensate world, and especially of the sociocultural world, in such a way as to reform it along the lines of the spiritual reality and of the ends chosen as the main value" (Sorokin 1957: 27).

According to Valsiner (2017), who analyzed the forms of integrated cultural systems from the point of view of cultural psychology, the *Ascetic Ideationalism* form

---

[2] It should not be forgotten that his monumental work *Social & Cultural Dynamics*, before being published in a revised and abridged version in one volume by the author in 1957 under the title *Social & Cultural Dynamics. A Study of Change in Major Systems of Art, Truth, Ethics, Law and Social Relationships*, had previously been published in four volumes (Sorokin 1937a, b, c, 1941).

[3] Sorokin, resuming the titles of the four volumes originally published, distinguishes these areas into: Art, Truth, Ethics, Law, Social Relationships, War and Revolution.

is a tool to facilitate the psychological alignment of personal cultures to societal expectations. This model is applied by many religious systems to determine a framework of ideological control on the individual's actions (examples include fasting, sexual abstinence, flagellation, and self-flagellation). On the contrary, *Active Ideationalism* aims at actively changing the Self (oneself) by changing society.

In the Sensate mentality, the reality perceived trough the sense organs prevails, while the relationship between individual and society is instrumental: the sensate mentality "does not seek or believe in any supersensory reality; at the most, in its diluted form, it assumes an agnostic attitude toward the entire world beyond the senses. The Sensate reality is thought of as a Becoming, Process, Change, Flux, Evolution, Progress, Transformation. Its needs and aims are mainly physical, and maximum satisfaction is sought of these needs (Sorokin 1957: 27). The method of meeting the needs does not depend on an inner change of individuals—who, through their meaningful interactions, actually create culture—but rather on a modification and exploitation of the outside world. This is the common feature to all forms of the Sensate culture mentality. If we consider item (4) again (*the methods of satisfaction*), we can distinguish three main varieties of this type: (a) *Active Sensate Culture Mentality* (Active "Epicureans"), which seeks the satisfaction of its needs and ends mainly through an efficient modification, adjustment, adaptation, reconstruction of the external milieu; (b) *Passive Sensate Mentality* (Passive "Epicureans"), which is characterized by the attempt to fulfil physical needs and aims, neither through the inner modification of "self", nor through an efficient reconstruction of the external world. Satisfaction occurs through a parasitic exploitation and utilization of the external reality as the mere means for enjoying sensual pleasures ("Life is short", "Carpe Diem", etc.); and finally (c) *Cynical Sensate Mentality* (Cynical "Epicureans"), which in the attempt to achieve satisfaction for their needs adopts the technique of the "mask". Depending on the situation, they don or doff a mask based on the greater returns in their physical profit.

All other culture mentalities are a mixed-system that takes elements from the Ideational and Sensate systems combining them in a variety of ways. This system is called Idealistic and it is in turn divided into:

> "**A.** *Idealistic Culture Mentality*. This is the only form of the Mixed class which is—or at least appears to be—logically integrated. Quantitatively it represents a more or less balanced unification of Ideational and Sensate, with, however, a predominance of the Ideational elements. Qualitatively it synthesizes the premises of both types into one inwardly consistent and harmonious unity. [...]: in other words, it gives *suum cuique* to the Ideational and the Sensate. **B.** *Pseudo-Ideational Culture Mentality*. Another specific form of the Mixed type is the unintegrated, Pseudo-Ideational mentality. One might style it "subcultural" if the term culture were used to designate only a logically integrated system (Sorokin 1957: 28–29).

Examining the characteristics of the Ideational and Sensate culture types and their Idealistic mixed version is clearly a different problem from investigating how these mentalities and their characteristics are distributed in cultural realities. And something else entirely is the actual behaviour of individuals and groups with respect to these mentalities.

3 The Superorganic (or Cultural) Phenomenon as a Social Problem

When it comes to culture, then, we do not usually refer to conditions affecting the individual. We rather refer to meaningful interactions between individuals, which in everyday life are closely related to the aspects of personality, society, and culture (meaning, norms, and values). Sorokin repeated this concept in several of his works, such as *Social & Cultural Dynamics* (1957), *Integralism is My Philosophy* (1958), and *Society, Culture, and Personality* (1962).

As previously stated, there is no unified definition of culture or cultural system, or a single form of it. Nor there is a unique analysis approach. All the disciplines falling under the category of social sciences have dealt with this topic and each based its contribution on the paradigms and peculiarities characterizing it. The various definitions coined over the decades have not clarified the concept, because it presents a high degree of relativity due to its indissoluble bond with the social and individual aspects. The whole of Sorokin's reflection (1962) is based on the indissoluble link between these elements, which binds all aspects of culture (meaning, norms, and values) to those of personality and society. This highlights the reciprocity between the living world and the social system. It represents the core moment when the focus is not only on the individual as the recipient of decisions, but also on the individual as a "subject" and an active part in the decision-making process. In this way, we move from an approach tending to reduce the social system to mere instrumental aspects, to an approach that considers the overall interactions between individuals and all other important variables that together constitute society.

Thus, changes in people's attitudes towards everyday life events, as well as in their ways of dealing with them, result from the combination between the psychic component and the broader understanding of the context in which they act: the idea of culture evolves together with contexts and societies, since they are both based on cultural mentalities. These are constructed through the meaningful interactions of individuals in the course of everyday activities. This is how the conjunction between acting and culture—*habitus* (Bourdieu 1979)—takes place as a *lifestyle* (consumer preferences and choices are unified in different environments). There is a kind of negotiation between the expectations of the individual and those of the society-system, through which it is possible to achieve satisfactory results for both parties.

Sociocultural processes (superorganic phenomena) permeate the lives of individuals. Or rather, as Sorokin stated in the first pages *of Society, Culture, and Personality* (1962), superorganic phenomena develop and are based only on interacting human beings and on the world they build. These interactions, arising through the daily roles, can transform the superorganic phenomenon into a *social problem*. The latter results from the relationship between "fact" and "structure" (of personality, society, and culture). It is the result of an interpretation and as such it is

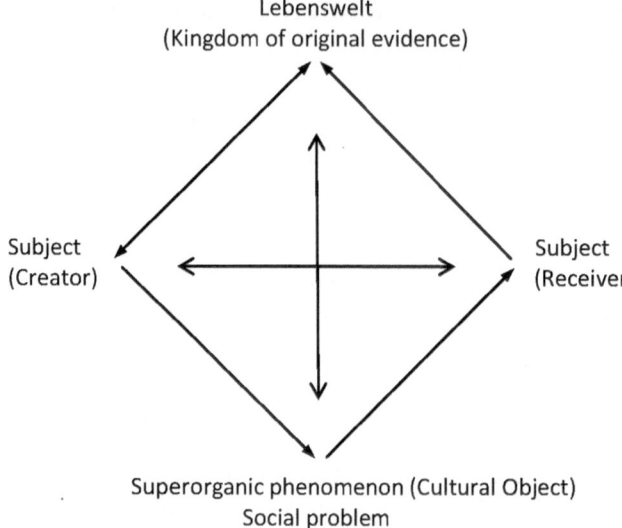

Lebenswelt
(Kingdom of original evidence)

Subject
(Creator)

Subject
(Receiver)

Superorganic phenomenon (Cultural Object)
Social problem

Fig. 5.1 Elements, connections and relationships involved in building superorganic phenomena as social problems

a cultural object[4] (Griswold 1994). Precisely because it is a cultural object, it can be interpreted—over time—as a socially defined social problem depending on the increase or decrease of shared forms of representation.

By adapting Griswold's "cultural diamond" (1994: 15) we can try building a model that illustrates the connections and relationships arising between the elements involved in the construction/production of the superorganic phenomenon as a cultural object in its form as social problem (Fig. 5.1).

Two aspects differentiate the model proposed here from Griswold's. The attribution of a "sense" to the connections (understood as direction), and the reference to the *Lebenswelt*, which Husserl (1970) defined as "kingdom of original evidence", rather than to the "social world". In other words, the starting point are common sense (built-in knowledge), and social interactions as preconditions for any reflection and insight into everyday life. Interaction takes the form of a reciprocal action with autonomous connotations that transcends those who carry it out. It is lies in a framework of meanings (culture) and it is both resource and constraint for the social system. From the *Lebenswelt* stem the meanings and representations, and from these all the other connections branch out. Considering Fig. 5.1 and moving counterclockwise, we can observe that the first connection is that with the creators of "sense" through a symbolic

---

[4]The *cultural object* "may be defined as shared significance embodied in form. In other words, it is a socially meaningful expression that is audible, visible, or tangible or that can be articulated. [...] Notice that the status of the cultural object results from an analytic decision that we make as observers; it is not built into the object itself" (Griswold 1994: 11). It should be noted that this definition, while clear for what concerns the modes and creators of cultural objects, does not make it clear that "products" differ depending on their source of production.

mediation allowing for the interpretation and construction of reality. Between *Lebenswelt* and individuals—understood as creators—there is a mutual connection that allows the other to be recognized. Creators can be individuals or collective subjects. Collective actions are those taken by a number of individuals who agree with each other and develop common strategies (groups). We disagree with the Durkheimian idea (Durkheim 1912) that cultural objects (culture system) are an exclusive product of society. Individuals, precisely because they are parts of a collective, are perpetually involved in meaning-building processes (through socialization and integration). Creators are in mutual connection with "receivers". Apart from the role they play in the process of cultural production (creator or receiver) or their number (individual or group), they are interacting with each other as parts of the same system, which has both similarities and differences. Through the manifestations of these processes, we reach the third element of the model, i.e. the superorganic phenomenon as a cultural object. Cultural systems, in general, transform events and objects into cultural objects by attributing to them a specific meaning that differs alongside culture. Similarly, some social phenomena are considered meaningful and transformed into cultural objects or, more specifically, they are transformed into social problems because they focus on the concerns of individuals and institutions. In other words, first identifying and then selecting social problems is crucial to the functioning and order of society, culture, and the processes of building the identity of individuals. This also explains the last mutual connection of the model, between the superorganic phenomenon (cultural object) and the lifeworld. Reciprocity is due to the fact that the lifeworld encapsulates all cultural objects. Their construction and production are influenced by the lifeworld as it connects individuals. We thus reach the fourth element of the model. Like all cultural objects, superorganic phenomena have their own "receivers"—individual and/or collective—that may or may not coincide with the "creators". Beyond their identification, "receivers" are not to be considered a "passive audience", as they are also producers of meanings referring to lifeworlds. At this point, the construction cycle of the superorganic phenomenon as a social problem comes full circle by returning to the *Lebenswelt*.

This adaptation of the "cultural diamond" is to be considered only a support tool. It has allowed us to elaborate an example of certain concepts and processes described by Sorokin concerning the indivisible sociocultural trinity (Sorokin 1948). This model has been useful in highlighting and expounding once again the relevance of personality—understood in Sorokin's sense as the subject of interaction (1962)—and the social reality that individuals experience in their everyday life in "building" and "producing" the "superorganic phenomena" and in "selecting" them as social problems.

References

Berger, P. L., & Luckmann, T. (1966). *The social construction of reality: A treatise in the sociology of knowledge*. New York: Penguin Books.

Bourdieu, P. (1979). *La distinction. Critique sociale du jugement*. Paris: Les Édition de Minuit.

Cuche, D. (1996). *La notion de culture dans les sciences sociales*. Paris: La Découverte.

Durkheim, É. (1912). *Les formes élémentaires de la vie religieuse*. Paris: Alcan.

Goody, J. (1986). *The logic of writing and the organization of society*. Cambridge: Cambridge University Press.

Griswold, W. (1994). *Cultures and societies in a changing world*. Thousand Oaks: Pine Forge Press.

Hall, S. (1997). The centrality of culture: Notes on the cultural revolutions of our time. In K. Thompson (Ed.), *Media e cultural regulation* (pp. 207–238). London: Sage.

Husserl, E. (1970). *The crisis of European sciences and transcendental phenomenology: An introduction to phenomenological philosophy*. Evanston: Northwestern University press.

Mangone, E. (2015). *Knowledge for the future of Mediterranean*. Areablu Edizioni: Cava de' Tirreni.

Ong, W. J. (1982). *Orality and literacy. The technologizing of the word*. New York: Routledge.

Sorokin, P. A. (1937a). *Social & cultural dynamics. Vol. I: Fluctuation of forms of art*. New York: American Book.

Sorokin, P. A. (1937b). *Social & cultural dynamics. Vol. II: Fluctuation of systems of truth, ethics, law*. New York: American Book.

Sorokin, P. A. (1937c). *Social & cultural dynamics. Vol. III: Fluctuation of systems of social relationships, war and revolution*. New York: American Book.

Sorokin, P. A. (1941). *Social & cultural dynamics. Vol. IV: Basic problems, principles and methods*. New York: Bedminster Press.

Sorokin, P. A. (1948). *The reconstruction of humanity*. Boston: The Bacon.

Sorokin, P. A. (1957). *Social & cultural dynamics. A study of change in major systems of art, truth, ethics, law and social relationships*. Boston: Porter Sargent.

Sorokin, P. A. (1958). Integralism is my philosophy. In W. Burnett (Ed.), *This is my philosophy. Twenty of the world's outstanding thinkers reveal the deepest meaning they have found in life* (pp. 180–189). London: George Allen & Unwin.

Sorokin, P. A. (1962). *Society, culture, and personality: Their structure and dynamics. A system of general sociology*. New York: Cooper Square.

Thompson, J. B. (1995). *The media and modernity. A social theory of the media*. Cambridge: Polity Press.

Valsiner, J. (2017). *Mente e culture: La psicologia come scienza dell'uomo*. Rome: Carocci editore.

Chapter 6
Personality and Human Conduct

1 The Personality as *Weltanschauung*

The term Personality refers to the subject of interaction—the third element in Sorokin's (1962) indivisible trinity (society, culture, and personality)—in the form of a single individual or of one or more groups. In both forms, the subject of interaction is the bearer of a *Weltanschauung* deriving from the dominant cultural mentality in the society (Ideational, Sensate, and Idealistic) in which she is born and lives: "individuals are the indispensable components of all social and cultural systems, their personalities (i.e.—the organization of their minds and behavior) obviously influence the framework of the social and cultural patterns" (Sorokin 1962: 342). This subject not only is physically within a culture, as she possesses a body, but she is fully part of it as a creator of meanings through her mind and soul: "the mentality of every person is a microcosm that reflects the cultural microcosm of his social surroundings" (Sorokin 1957: 606). The three components of personality, inseparable in Sorokin's theory (body, mind, and soul), represent the three means for knowledge (empirical-sensity, reason, and intuition). They are the very essence of personality. In this dialectical process, a direct relation is established between the dominant social mentality and the conduct of the subject within it. However, this relation does not result in a one-to-one correspondence. It is very variable in some societies and more explicit in others. This creates a seamless link between mentality and conduct: "there is hardly any clear-cut boundary line between mentality and behavior. They imperceptibly merge into each other, and many phenomena of mentality are at the same time phenomena of conduct and behavior, and vice versa" (Sorokin 1957: 608–609). It is therefore clear that the superorganic aspects of personality are thus clearly *not* determined by birth (i.e. biologically). They are shaped by the social and cultural environment of belonging that also characterizes the person's sociality. Indeed, all her expressions (from beliefs to emotions) and meaningful actions occur and are processed within the aggregates where her interactions take place. The subject identifies herself in the social system, and is in turn socially

© The Author(s) 2018
E. Mangone, *Social and Cultural Dynamics*, SpringerBriefs in Psychology,
https://doi.org/10.1007/978-3-319-68309-6_6

identified (Cuche 1996), through a series of categories (sex, age, class, nationality, etc.). Unlike in the recent past, these processes no longer follow regular stages. Today, even more than in the past, the regular progression of people's lives is depleted due to increased social uncertainty. This does not allow the allocation— and thus the clear membership—of individuals to a single social circle (Bauman 2001) within which to occupy a well-defined position that represents the starting point for all other positions. In other words, the personality of the contemporary subject arises at the intersection of multiple social circles. There is no single symbolic universe, rather, there is the simultaneous presence of more *provinces of meaning* (Schütz 1932). The simultaneous presence of different definitions of the same reality (different *Weltanschauung*) and the multiplication of *Lebenswelt* place the individual in contact with barely integrated and often dissonant systems of meaning (Festinger 1962). This entails flexibility and differentiation to counter the conflicting aspects of the different realities. Once again, the inseparability of individual, social, and cultural is confirmed, also because of the following conditions:

"(*a*) the concomitant development of the individual mind and the social structure; (*b*) by the close correspondence between the structure of the individual's egos and the structure of the groups to which the individual belongs; (*c*) by the effective determination of the individual's conduct by the groups in which he lives; (*d*) by the general defining of the content of the individual's mind (his scientific, philosophical, religious, ethical juridical, aesthetic and other values) by the cultural world in which he moves" (Sorokin 1962: 343).

At this point, any thoughts that individual development is related to sociocultural organization are now gone. Sorokin himself, in *Society, Culture, and Personality* (1962: 344), identifies some peculiarities of individual development (he identifies seven of them to be precise), of which we propose here a non-exhaustive but exemplifying synthesis: (1) Without human interaction, individual development is impossible, and without the transmission of experience between generations, no accumulation of knowledge would have been possible, as it is not biologically transmitted; (2) Without a collective experience in continuous transformation, it would not be possible to distinguish good from evil, just from unjust, and normal from pathological; (3) Without a context of meaningful interaction (typical of the individual), the evolution of individual faculties (memory, categorization, etc.) and certain social categories (membership, identity, difference, etc.) would not be possible. Nor would have been possible the development of language based on the transformation of society—think of the evolution of language with the advent of the electronic era.

This evidence does "not leave doubt as to the correlation of growth, variation, and decline of social structures and cultural systems on the one hand, and the development and changes of the human personality, on the other hand" (Sorokin 1962: 345): two sides of the same coin (sociocultural reality). It follows also another aspect of the relation between personality and social reality. The former is not merely the result of the conflict between conscious and unconscious (as in Freud's theory), but instead the result of a more complex negotiation between several aspects. In this regard, Sorokin describes, in general terms, four levels whose combination leads to the development of the acting personality: (1) the *biological unconscious*, which is the lowest level and represents instincts; (2) the *biological conscious* or the *bioconscious*, the level in which the energy of the person's body becomes conscious and is able to control instincts; (3) the *sociocultural conscious* or *socioconscious*, which is formed by the ways of interacting with others, by the personal experiences that are realized together with others through collective life and that are transmitted from individual to individual, from group to group, and from generation to generation; and (4) "The *supraconscious* level represents the fourth and highest level of psychic activities, mental energies, and personality. It is the level of inspiration for spontaneity, originality, and creativity of genius. In the writings of all civilizations it is designated by numerous expressions, such as "the divine element in man", "the manifestation of divinity", "the sublime energy of Truth, Beauty, and Goodness", "creative genius" (Vexliard 1963: 167). At this point, however, if we consider personality—as described by Sorokin—and reasoning by analogy, the concept of identity[1] (Mead 1934), we may note that perhaps our author has forgotten to point out that personality is expressed in two ways: the person's *doings* and her *being*. On these two ways of expression, we will briefly refer to two other scholars. Ricoeur (1990), who distinguishes the *idem* identity, referring to the social definition of the individual and her continuity (*mêmeté*), and the *ipse* identity, related to the unpredictability of single individuals; and Mead (1934) who identifies the *I* (analogous to the *ipse* identity), which recognizes itself only on the historical level as the continuity of life moments, and the *Me* (similar to the *idem* identity), which represents the objectivated set of social patterns that an Individual interiorizes, bringing them to unity in the *Self*.

1.1 Identity in Contemporary Society

In contemporary society, these processes become more complex because the social space where the individual is located becomes increasingly widespread day after day, thanks also to the mass media that "opened up" and "multiplied" the lifeworlds. The horizons of understanding have expanded: they are no longer limited to face-to-face interaction; instead, they are shaped by mediated communication networks. The excessive mingling of *actual* and *virtual* life means that Self and reality are in perpetual conflict, denied and reorganized—if denial has not already produced the physical death of the individual. It may be interesting to introduce here Goffman's contribution (1968). The author, based on these dynamics, distinguished individual identity into *personal and social*. The social identity was in turn distinguished into *virtual* (ascribed to the individual based on her appearance, i.e. of what others imagine and following which we can reach only a mere approximate and supposed), and

---

[1] Identity is made up of the peculiarities of the individual personality, which make it possible for her to position herself in the social environment in relation to the other subjects who are co-present in this environment. It is a system of representations on which everyone bases: (a) her existence as a person; (b) recognition and acceptance by others, the group and the culture of belonging.

actual (which determines with a higher degree of certainty to which category that individual belongs). The social identity's constant oscillation between *virtual* and *actual*—due to the constant change of situation—forces individuals to redefine their identity (personality). This constant oscillation drives individuals to adopt all the necessary strategies to maintain ties with the outside world, *which they must keep into account* when taking action. This situation causes a progressive de-structuring of the Self, because there is less and less chance of using the strategies with which individuals usually try to escape from the dynamics of power.

In the event of conflict of positions—negotiation between rights and duties, or between social expectations and individual desires—this plurality of identities creates antagonism (and inner conflict); conversely, in case of positive negotiation it leads to equilibrium (inner harmony). Obviously, people's actions result from these inner situations. The former can be antagonistic or solidaristic, depending on the prevailing position. Therefore, Sorokin comes to the following first conclusions about personality:

> "(1) that the *forms* or *patterns* of almost all the overt actions and reactions (or conduct and behavior) of the members of each dominant type of culture are shaped and conditioned by it; (2) that each culture, to some extent, stimulates many activities and inhibits many others in conformity with its nature; (3) that only the actions and reactions that are most closely related to the elementary biological needs experience, in conformity with our first proposition, a comparatively mild conditioning by the dominant culture, *so far as the performance or nonperformance, the frequency, and the intensity of their satisfaction are concerned.* The forms in which these activities are discharged are also conditioned by the dominant culture" (Sorokin 1957: 609–610).

It is a process whereby we can assert that the person is the fruit of her experiences. The interacting individual (personality) stems from her daily actions, which neither casually accumulate nor create an intricate system of notions, impressions, and sensations. These experiences are selected, organized, recalled in a specific order and according to a *fil rouge* of inner consistency so as to form the personality of the individual (identity). Identity building takes place over time through processes of differentiation and integration, making it subject to change due to the very interactions experienced by the person in her daily life. The individual is characterized by the fact that she is a unique entity, performing a particular task, which is different from that of other entities (differentiation), but which is also part of a more general task (integration) concerning her social life.

According to Sorokin, these transformations should aim at changing the personalities by guiding them towards a more altruistic approach. To do so, one has to focus on the potentialities of the higher levels of the personality (supra-conscious and conscious), able to control the lower one (unconscious). "The transformation of the individual implies a triple action: (a) the reorganization of the different egos by placing them under the authority of a vigorous supraconscious and conscious, which are developed to the maximum; (b) the revision of all values and their subordination to the highest, creative values (love, truth, and beauty); (c) an affiliation of the individual with those groups which cultivate these positive values, and his break with those groups which appear egoistic and negative" (Vexliard 1963: 174). However, being oriented towards altruism does not mean denying the value of individuals, but rather building bonds of solidarity and justice between them. Starting from self-knowledge, the individuals must tend towards the knowledge of action (what they ought to Do) and of essence (what it should Be): this is the emblematic virtue of human beings. The *Weltanschauung* derives from the constant search by individuals for a combination between being and doing; experience, or research, are two ways of reading the reality of the world and, each in its own way, they both aim to attain the well-being of the individual in society.

2 Human Conduct and Uncertainty

Concerning what Sorokin argues about personality, we now wish to propose a more current reading of some of the concepts he dealt with until the 1950s. Reflecting and arguing on contemporary society leads us to broaden our perspective in order to sketch an overview of how all the elements (personality, society, and culture) involved in the different processes are interconnected and how they permeate their own changes.

Contemporary sociological reflection has highlighted that discussing an individual means talking about her actions embedded in a cultural and social context, as well as in an interaction context. Sorokin himself—although living in a very different historical moment—has never denied that personality derives from the individual in interaction with others and with the outside world. Signification processes vary together with the *umwelt*, where the latter term is intended in Schütz's acception (1932) as "current social context" and not as understood by Uexküll's ethology (1992, original ed., 1934). If we relate these dynamics also to the individual performing the action, we can see that for her the same action can assume different meanings depending on the temporal arc in which it is placed (temporal perspective). Before proceeding with the description of this latter aspect, however, we should point out that Sorokin, in the book *Sociocultural Causality, Space, Time* (1964) had given much importance to this variable, and also to that of space. In this book, the author distinguishes time into quantitative and qualitative: "in our modern society, side by side with quantitative time (which itself is in a degree a social convention), there exist a full-blooded sociocultural time as a convention, with all its 'earmarks': it is qualitative [...]; it is determined by social conditions, and reflects the rhythms and pulsations of the social life of a given group" (Sorokin 1964: 197). This subdivision is made even clearer by resuming Schütz's arguments (1932). Actions can have a meaning and a representation before they are actually carried out (future project, *sense of production*), while they are being carried out (current experience, *sense of produced*), and, finally, after they have been accomplished (past memory, *self-understanding* and *understanding others*). In simpler terms, we can state that, following Schütz's theory, assigning meaning to an action is arbitrary because it is linked to a project built by the individual, and thus susceptible to modifications. The action planned and the one actually undertaken often do not coincide, and this implies the necessary distinction between the meaning of the action planned (end) and the meaning of the action completed (cause). This implies that actions can be understood in their significance only if we understand their purpose and the temporal arc to which they relate.

People's actions in contemporary society, with its global characteristics, are closely related to the concept of uncertainty. An inevitable human condition, as Pascal (1901) claimed some centuries ago, which considers human beings as sailing in a vast sea, pushed from one extremity to the other but always uncertain and fluctuating. Human beings, since their earliest forms of organization, have tried to find sources of knowledge that would allow them to expand their degree of certainty (understood as security). However, this need is never satisfied because human beings, despite their efforts, are unable to gain complete knowledge of a certain situation, and this obviously raises the degree of uncertainty.

This oscillation between search for certainty and increase in uncertainty—also due to the higher complexity of contemporary society—characterizes what Bauman called the "society of uncertainty": "Many a feature of contemporary living contributes to the overwhelming feeling of uncertainty: to the view of the future of the 'world as such' and the 'world within reach' as essentially undecidable, uncontrollable and hence frightening, and of the gnawing doubt whether the present contextual constants of action will remain constant long enough to enable reasonable calculation of its effects" (Bauman 1997: 21). For human beings, taking a position of control has always meant trying to improve their living environment by reducing uncertainty and crisis situations.

In *The Reconstruction of Humanity* (1948), Sorokin tries precisely to sketch how humanity can emerge from the highly uncertain situation following World War II, which has in fact led to a catastrophic crisis. According to the author, the only possible way out from this crisis is altruism, that becomes the only tool for peace and survival.

Human conduct, however, is not always positively oriented towards the other, a condition that strengthens the degree of uncertainty and influences future actions. Diversity is considered a constraint to everyday life and a threat to the future rather than a resource. These uncertain conditions play a paramount role in the construction process of the social representations of uncertainty and hence of the future project of people's lives. If expectations of trust do not replace the lack or excess of information in the subjects (reduction in uncertainty) there is no positive reassurance against contingent actions or situations.

3 Human Conduct, Trust, and Risk

If uncertainty causes people's lack of control over their lives, they will tend to survive by looking for ways to reduce it so that decisions can be taken in a more conscious way. Individuals aim to convert uncertainty into certainty by adapting their behaviour to social norms, habits, and customs, by following an order from a recognized authority, or by adopting contractual action patterns. And if that is not possible, individuals will at least tend to turn uncertainty into a calculable risk (Barbieri and Mangone 2015). However, reducing uncertainty on the basis of trust relations appears an almost paradoxical practice, as these relations entail a compulsory risk: encountering the other and, moreover, counting on her positive action in order to be able to grant her our trust[2] (Gambetta 1988; Fukuyama 1995).

We advance here Sorokin's concept of *creative altruistic love*, first introduced in *The Reconstruction of Humanity* (1948), and fully developed and discussed in *Altruistic Love* (1950) and *The Ways and Power of Love* (1954). This seems, in Sorokin's eyes, the only way to save humanity. A peaceful, harmonious, and creative society can only exist when its members possess love, sympathy, and compassion, ensuring mutual aid, cooperation, and fair treatment for all its members. Under these conditions, the members of the society are united in a collective "we" that leaves no-one isolated because everyone is an essential part of a creative community.

All these elements ultimately lead to two aspects: reciprocity and belonging. The latter term, however, is ambivalent: on the one hand, individuals develop it for their own benefit, on the other, the very community within which this concept is developed uses it in order to continue to "exist".

Up to this point, we neglected the personality as a situational and meaningful act of individuals (positive or negative orientation towards the object of representation). Superorganic phenomena, as Sorokin intends them (1962), provide a social representation of reality. And this is defined by the relation between the person's complex cognitive systems and the symbolic relations systems between all social actors (individual or collective). In the flow of everyday experience, people try to articulate a dialogue with society within the existing symbolic relations between individuals, groups, and institutions. Representations, as symbolic constructions influenced by the social position of the actors that produce them (Jodelet 1984), take on the function of "conventionalizing" objects, individuals, and sociocultural phenomena. They give them a precise form, assigning them to a category and confining them to a model, distinct and shared by a group of individuals. The action becomes therefore the final event that combines all these aspects; however, in order to reflect on actions, one must look at individual action starting from Max Weber's idea. According to Weber, sociology is the comprehensive science of social action (unlike Durkheim, who claimed the prevalence of the structure on the individual). Weber's *Verstehen* (Sociology of Understanding) (1978) qualified human action as a social action only when "it has meaning", that is, it is guided by motivation. In other words, action is social when individuals in their actions take in the account of the action of others

---

[2] Trust—in general terms—is defined as an expectation of positive experiences, matured in conditions of uncertainty and in the presence of a cognitive and/or emotional load so that it allows to overcome the threshold of hope. Specifically, trust-based interaction—both institutional and inter-individual—is closely related to positive or negative experiences (aspects related to "meaning"), to the "conditions of uncertainty" and to the "cognitive and/or emotional burden" of individuals. Social interactions, however, include in their daily occurrence both relations with each other and relations with institutions.

(independently from the their presence or absence). The action qualifies as social because it always refers to the behaviour of others, which influences its evolution. Social action must be defined in terms of "objective meanings" of the individual's activity. It is the key to reading modern Western society, which, according to Weber, becomes increasingly dominated by a goal-instrumental rationality.[3] If this is the key to read society, we must forward some clarifications. In light of Sorokin's previous descriptions, the actions of individuals appear dominated by a logic based on the search for balance between "goals" and "means".[4] This starts with the selection of "goals" on a hierarchical scale referable to Maslow's (1954), where the motivation for an action arises from the universal tendency to satisfy certain orders of needs, which are different by nature and complexity, in ways other than according to a logic of purpose-oriented rationality. In this way, actions cannot be enveloped in an ideal model, as they escape from all rational logics.[5]

In this context, uncertainties are exasperated for two reasons: on the one hand, there is an almost total absence of rules on the legitimate procedures for achieving the "goals"; on the other hand, cultural goals are proposed at all social levels without there being a real opening to all the institutionalized ways to reach them. Negative conducts are therefore favoured by these anomic conditions, and take on different forms, which differ according to how the antinomy between the "goals" posed by culture and the "means" used to achieve them is resolved. Resolutions cannot be observed merely and exclusively as subjective responses elaborated to cope with individual discomfort (subjective anomie). They must be seen as collectively and consciously developed responses based on real social contradictions, in which time becomes a strong variable that forces individuals to make predictions for the future over choices made in the past.

To define, within the limits of Sorokin's theoretical speculations, what we outlined here on human conduct, we must trace everything back to a banal but brilliant intuition by this scholar. According to Sorokin, the future of humanity and its

---

[3] We should point out that Weber defines a typology of social action through the conceptual tool of the *ideal type*, distinguishing it into: instrumentally rational action, value-rational action, affectual action, and traditional action.

[4] The equilibrium processes between "goals" and "means" have also been taken up in Merton's theory of deviance (1968)—for a time, the author was also Sorokin's student at Harvard. In summary, according to Merton, contemporary societies are characterized by the importance attributed to the "cultural goals" and the importance placed on the "means" to reach them. This creates a dissociation between the terminal values and the instrumental values. This is particularly the case when there is an attenuation of the importance of legitimate means in favour of the use of any effective means of achieving the cultural goals. When the legitimate practices for reaching a cultural goal are shadowed, the individual faces a form of "anomie", to which she remedies through actions patterns that vary depending on her position in the social organization.

[5] An example in this sense is those who adopt life-endangering lifestyles (driving at speed, use/abuse of psychoactive substances, etc.). According to Lyng (1990) these are *edgework*, i.e. actions in which the individual renegotiates the boundary between life and death. They are the satisfaction of the higher-placed need in the hierarchical scale, that is, self-realization (the need to fully realize their potentialities and expectations on their own existence). Again, in this case we see the search for a balance between "goals" and "means".

development is in the hands of humanity itself and not of other subjects or institutions. This arduous task must not be attributed to law or education, nor to religion or economy, or science, even though the latter has a very precise role in accompanying the processes of improving the lives of individuals and communities. This arduous task is for humanity, and no others.

References

Barbieri, A. S. A., & Mangone, E. (2015). *Il rischio tra fascinazione e precauzione*. Milan: FrancoAngeli.

Bauman, Z. (1997). The making and unmaking of strangers. In *Postmodernity and its discontents* (pp. 17–34). Oxford: Blackwell.

Bauman, Z. (2001). *The individualized society*. Cambridge: Polity Press.

Cuche, D. (1996). *La notion de culture dans les sciences sociales*. Paris: La Découverte.

Festinger, L. (1962). *A theory of cognitive dissonance*. Stanford: Stanford University Press.

Fukuyama, F. (1995). *Trust*. New York: The Free Press.

Gambetta, D. (Ed.). (1988). *Trust: Making and breaking cooperative relations*. New York: Basic Blackwell.

Goffman, E. (1968). *Stigma. Notes on the management of spoiled identity*. Harmondsworth: Pelican Books.

Jodelet, D. (1984). Représentation sociale: Phénomènes, concept et théorie. In S. Moscovici (Ed.), *Psychologie sociale* (pp. 361–382). Paris: PUF.

Lyng, S. G. (1990). Edgework: A social psychological analysis of voluntary risk taking. *American Journal of Sociology, 95*(4), 851–886.

Maslow, A. H. (1954). *Motivation and personality*. New York: Harper & Row.

Mead, G. H. (1934). *Mind, self & society from the standpoint of a social behaviorist*. Chicago: Chicago University Press.

Merton, R. K. (1968). *Social theory and social structure*. New York: The Free Press.

Pascal, B. (1901). *The thoughts of Blaise Pascal*. London: George Bell, originally published in 1670.

Ricoeur, P. (1990). *Soi-même comme un autre*. Paris: Seuil.

Schütz, A. (1932). *Der Sinnhafte Aufbau der sozialen Welt*. Vien: Springer.

Sorokin, P. A. (1948). *The reconstruction of humanity*. Boston: The Bacon.

Sorokin, P. A. (1950). *Altruistic love: A study of American good neighbors and christian saints*. Boston: Beacon.

Sorokin, P. A. (1954). *The ways and power of love. Types, factors and techniques of moral transformation*. Boston: Beacon.

Sorokin, P. A. (1957). *Social & cultural dynamics. A study of change in major systems of art, truth, ethics, law and social relationships*. Boston: Porter Sargent.

Sorokin, P. A. (1962). *Society, culture, and personality: Their structure and dynamics. A system of general sociology*. New York: Cooper Square.

Sorokin, P. A. (1964). *Sociocultural causality, space, time. A study of referential principles of sociology and social science*. New York: Russel & Russell.

von Uexküll, J. (1992). A stroll through the worlds of animals and men: A picture book of invisible worlds. *Semiotica, 89*(4), 319–391. originally published in 1934.

Vexliard, A. (1963). Sorokin's psychological theories. In P. J. Allen (Ed.), *Pitirim A. Sorokin in review* (pp. 160–188). Durham: Duke University Press.

Weber, M. (1978). *Economy and society: An outline of interpretive sociology*. Berkeley: University of California Press.

Chapter 7
From Creative Altruistic Love to the Ethics of Responsibility

1 The Creative Altruistic Love in Sorokin

"At the present juncture of human history, a notable increase of an unselfish, creative love (goodness) in the superorganic world is the paramount need of humanity" (Sorokin 1958: 184). This is the end point of Sorokin's strenuous work. In fact, beyond any form of "prophetic vision" and all the controversies that followed the above statement, the Russian-American sociologist had an "intuition" synthesizable in the idea that humanity itself must act for its own salvation. Hyper-individualism has led to conflicts between individuals and groups, whose negative effects reverberate on these same individuals and groups:

> "In the twentieth century interhuman strife assumed the catastrophic proportions of two world wars and many other wars, of endless bloody revolutions and revolts, not to mention crimes and milder forms of the "struggle for existence". At present, due to the discovery of the intra-atomic secrets and to the invention of Apocalyptic means of destruction, this moral anarchy begins to threaten the survival of mankind and especially the continuation of its creative mission. The situation explains why a notable increase of unselfish, creative love in the total human universe is the paramount present need of humanity" (Sorokin 1958: 185).

The American academic world's criticism of Sorokin's ideas was probably caused by an incomplete or superficial knowledge of both his previous works and those in which the author condenses the discussion of creative altruistic love (Sorokin 1948, 1950, 1954). And, most likely, his activities at *The Harvard Research Center in Creative Altruism*[1] were also little known. Therefore, in order to avoid the same problem, here one chooses to clarify some of the key aspects of what appears to be the only means of saving humanity.

---

[1] The centre was established in 1949 thanks to founding by Mr. Eli Lilly and the Lilly Endowment, with the aim of promoting interdisciplinary research and organizing symposia on altruism, analyzing its types, aspects, and dimensions, as well as its effects on the individual, social, and biological life (health and well-being).

© The Author(s) 2018
E. Mangone, *Social and Cultural Dynamics*, SpringerBriefs in Psychology,
https://doi.org/10.1007/978-3-319-68309-6_7

"A peaceful, harmonious, and creative society can exist only when its members possess at least a minimum of love, sympathy, and compassion ensuring mutual aid, co-operation, and fair treatment. Under these conditions its members are united in one collective 'we' in which the joys and sorrows of one member are shared by others. In such a group a member is not an isolated 'atom', but a vital part of a creative community [...]. Exercise your legal right and perform your legal duties when they do not harm anyone else and when they do not violate the rights and duties of others—such is the essence of marginal altruism, slightly above the purely legal conduct prescribed (Sorokin 1948: 57–58).

Sorokin's attention never deviates from what he described as the indivisible sociocultural trinity (personality, society, and culture). At this point in his studies, however, his attention focuses on the individual and her mentality. Moreover, Sorokin never departs from his idea of sociology as a science engaged in the study of meaningful interactions between all the elements of superorganic phenomena. A discipline able to show the way for improving the living conditions of individuals. And here are the origins of another of Sorokin's criticisms towards the *modus operandi* of some social sciences (particularly sociology and psychology), i.e. being "negativistic". These disciplines are able to reveal only negative or pathological phenomena, without ever pointing to positive and healthy ones.

According to Sorokin, altruistic love is not just a feeling, but a positively vital force that can push phenomena towards the highest levels of solidarity (social interaction). In light of this, the terms used by Sorokin since the book *Contemporary Sociological Theories* (1928) to qualify the conduct of human beings are not "conflicting" and "cooperative": with a telling choice, he adopts the terms "antagonistic" (or "compulsory") and "solidaristic". The latter term is not by chance: it is precisely the social responsibility of solidarity that is entrusted with guaranteeing the safeguard of social vulnerabilities, thus presupposing reciprocity. The main problem of a constantly changing society is the lack of mechanical solidarity ties—as *per* Durkheim. The person's action emerges as a causal dependence between her physical involvement and the pressure exerted on her by the environment. The term "solidarity" therefore presupposes a greater involvement of all the interacting parties in the social system. In this way, not only we avoid neglecting social protections for the more vulnerable people, but we also stimulate individual energies and autonomous initiatives to strengthen the protection and safeguard for all people. Applying a solidarity model therefore leads to two important transformations: on the one hand, institutions must assume a control function by guaranteeing individual freedoms and offering a minimum universalistic protection; on the other hand, there is the multiplication and differentiation of the individuals involved in the decision-making processes related to collective well-being goals.

The idea of love fits into this theoretical framework as "the supreme and vital form of human relationship" (love relationship) and as such the ways, forms, and power of this energy (love energy) are to be studied. This force is likened, for their similarities, to an iceberg: "Love is like an iceberg: only a small part of it is visible, and even this visible part is little known. Still less known is love's transempirical part, its religious and ontological forms. For the reasons subsequently given, love appears to be a universe inexhaustible qualitatively and quantitatively. Of its many forms of being the following can be differentiated: religious, ethical, ontological, physical, biological, psychological, and social" (Sorokin 1950: 3). These forms actually refer to the very aspects of love: (a) religious love, refers to the experience of love for God or the Absolute; (b) ethical love, *"is identified with goodness itself. Love is viewed as the essence of goodness inseparable from truth and beauty"* (Sorokin 1950: 6); (c) ontological love, is considered the highest form of unifying, integrating, and harmonizing creative power or energy. This is the "core" of love, because it makes the world go round and without it we would witness the collapse of the physical, biological, and social world (D'Ambrosio et al. 2014); (d) physical love, refers to love expressed through the unifying, integrative, and ordinating energies of the universe; (e) "The biological counterpart of love energy manifests itself in the very nature and basic processes of life. This energy, still little known, and often called the 'vital energy' that mysteriously unites various inorganic energies into a startling unity of a living—unicellular or multicellular—organism" (Sorokin 1950: 9); (f) psychological love includes all the intellectual aspects of emotional, affective and desire experiences. For its very nature, psychological love is an "altruistic" experience; (g) social love is the last of the forms identified by Sorokin "on the social plane love is meaningful interaction—or relationship—between two or more persons where the aspirations and aims of one person are shared and helped in their realization by other persons" (1950: 13). It follows that love not only has many aspects and forms, but it also has various dimensions, of which Sorokin identifies five: intensity, extensity, purity, adequacy, and duration.[2]

Considering love as an energy, akin to any other type of energy, means that it can be produced, accumulated, and distributed by individuals and institutions. These are the peculiarities that make creative altruistic love a powerful tool for the reconstruction of humanity, which was falling into a marked sensualism due to the transformations of its cultural mentality. At the time when Sorokin expressed his "foolish" ideas—as they were then defined by some of his colleagues (Sorokin 1955)—and still today while this work is coming into being, no solution has yet been found for the devastations and wars. This is because individuals tried to act from the outside, thinking of changing political and economic institutions without intervening on individuals. These attempts are destined to fail because, to change institutions and the economic system, it is necessary to change the individuals acting in these very institutions and systems. Or, rather, it is necessary to transform people's way of interacting by orienting them towards the love relationship that characterizes a free, harmonious, humanistic, and creative society.

Sorokin's cannot however be termed "sociological humanism"; instead, we could speak of a committed "humanistic sociology". A sociology that does not analyze and study only social phenomena, but a science that, with its peculiarities,

---

[2] On these five dimensions, some scholars (Levin and Kaplan 2010) developed and validated a measure of love, the Sorokin Multi-dimensional Inventory of Love Experience (SMILE). In view of Sorokin's opposition to quantophreny and tests, leading to an excessive simplification of social reality, I doubt that he would have been particularly enthusiastic about this application of his theoretical framework on love.

contributes to the analysis and study of the most human part of individuals and society (living man). The purpose is to contribute to the discovery of human beings as creative and responsible social actors.

2   Beyond Sorokin … the Ethic of Responsibility

According to Sorokin, change must start from the rediscovery of the positive values of man, and science becomes a guide to this end also by overcoming strictly sensuous knowledge models. It is not merely a sociology of the crisis: it is a "critical sociology" not limited to the analysis of the degenerative processes of society, but that seeks their deep roots by denouncing the negative factors determining them. Applying these assumptions implies understanding the mechanisms through which human beings make their own decisions. These dynamics highlight the issue of the *choice*. The latter, in turn, must *take different forms depending on* the temporal dimension and degree of knowledge of situations, as well as to who and how takes decisions (individuals or their representatives—politicians).

Who decides (individuals or politics) bases her decisions on the cultural mentalities and her degree of knowledge of a given situation. I am back to the problem that Sorokin raised in the *Integralism is My Philosophy* (1958) on the construction of an integrated system of knowledge that may hold together the three forms: empirical-sensory, reason, and intuition. A knowledge system able to provide as many elements as possible to understand superorganic phenomena, so as to have the opportunity, whenever possible, to foresee their transformation. The dynamics described here prompt us to state the desirability of the reconstruction of "humanity". A new humanity that can no longer be intended only as a right, but also as the duty to act on the basis of the ethics of responsibility. And this applies to individuals, politics, and institutions.

People's choices, however, beget two paradoxes (Melucci 1994): the first is that individuals are *forced* to choose. And having to choose means having to face responsibilities (based on the abilities and rights that an individual has or believes to save, and her degrees of freedom); the second is that the individual is obliged to select the range of possible alternatives at that specific time (in the future, the alternatives might change) and on which she will pour out the energies to achieve her purpose. If, at the dawn of our species, the influence of the environment prevailed over that of human beings, today we experience the opposite situation. The individual tries to prevail over the environment and deeply changes its conditions through techniques and technologies that do not always produce positive effects. The principle of responsibility (Jonas 1984)—akin to Sorokin's idea of altruistic love—aims at being the ethical foundation of actions to preserve both the human being and the integrity of his world for future generations. "If each of us, imbued with a deep sense of responsibility, 'watches his step', avoiding the selfish abuse of his functions, most of our social problems can be easily solved and most catastrophes prevented. On the other hand, without effortful self-education in altruism on the part of every individual, no social transformation is possible" (Sorokin 1948: 234).

Ethics—understood as the objective and rational foundations that make it possible to distinguish human conduct into good, just, or morally lawful, with respect to conducts considered to be bad or morally inappropriate—has characterized human life since ancient times. These principles—or their pursuit, usually distinguished in the two main dimensions of secular and religious (particularly Christian)—allow individuals to manage their own freedom. Freedom exercised within its limits, without violating the right to existence of other individuals. Today, when speaking of ethics, we are most likely to repeat what has already been said by many other scholars, but we cannot avoid this danger in addressing the issues of the development of humanity.

The contrast is usually between the two classic dimensions of ethics (secular and religious), but to overcome it we should perhaps speak of a third dimension: the public one, or that of responsibility. Concerning the first point, however, I can say that there are no real secular and religious values. This leads to the assertion that opposing secular ethics to religious ones over issues related to the development of humanity would be superficial and sterile. For this reason, I will advance here no speculations or guesses based on the opposition of these two dimensions of ethics, as it would result in a vague, worthless framework, and hence the need for a public ethics and particularly an ethics of responsibility. This is based neither on individual morality nor on collective ethics (secular or religious), but instead represents the *Weltanschauung* that is being transformed. For example, if we think about the technology that men use to suppress nature,[3] it is unthinkable that the old rules of "neighbourly" ethics can continue to be valid (justice, mercy, honesty, etc.). These rules must be reassessed and amended, due to the growing sphere of collective action, in which actors, actions, and effects are no longer the same as before. Man, in the age of technique (Gehlen 1989), has changed the meanings of actions, the affected objects, and the following consequences. No form of ethics in the past took into account the global condition of human life, the distant future, and the survival of the species. The very fact that these factors are at stake today sets the need for a new conception of rights and duties, for which traditional ethics does not offer principles and/or a doctrine.

The ethics of responsibility is neither objectivism nor hyper-subjectivism. It is looking for intersubjective and intercultural values that help dialogue between the different positions by orienting them towards collective good.

> "An imperative responding to the new type of human action and addressed to the new type of agency that operates it might run thus 'Act so that the effects of your action are compatible with the permanence of genuine human life'; or expressed negatively: ' Act so that the effects of your action are not destructive of the future possibility of such life'; or simply: 'Do not compromise the conditions for an indefinite continuation of humanity on earth'; or, again turned positive: 'In your present choices, include the future wholeness of Man among the objects of your will'" (Jonas 1984: 11).

---

[3] The nature has been particularly vulnerable to (more or less irreparable) damage, which also affects the well-being of humanity and, above all, future generations (pollution, deforestation, greenhouse effect, etc.).

In light of these considerations, individuals, irrespective of their religious beliefs and political affiliation, can orient themselves towards an ethics of responsibility that would guarantee collective good. The real problem lies in the fact that any moral rule has its exceptions, and this determines the need to identify the dominant rule among the conflicting ones. A new contrast would thus emerge between the Kantian principle of "never to use other people merely as means to an end, but always *also* as ends", and the utilitarian idea spurring people to always choose "actions that maximize their utility and happiness". Weber has well interpreted this conflict (Lassman and Speirs 1994; Lassman et al. 1989): plurality of values arises in the form of dualism between the ethics of moral convictions (*Gesinnungsethik*—also called of intentions or principles), and the ethics of responsibility (*Verantwortungsethik*). The former refers to absolute principles, which are adopted irrespectively of their consequences (e.g. religious ethics); the latter, instead, to all those cases where special attention is paid to the relation between means and ends, and the consequences of the action.

In summary, following the ethics of responsibility means that every individual must bear the consequences of her actions (for good and for evil) toward herself and others (beyond the proximity of time and space). In light of these reflections, epitomizing human conduct in one single general principle would be an ill-fated, absurd approach. The idea of responsibility for one's choices in contemporary society has often been deferred to law—not as ethics of responsibility, where there are no secular people and Catholics, believers and non-believers, but simple individuals—but as ethics reduced to simple procedure. If we ignore law-related aspects, the issues on the construction of a novel humanity could be exposed in terms of a negotiation between individual freedoms and responsible freedoms. In this way, the drive to self-realization cannot be conceived without the commitment to others and to the community in the broad sense. Assuming that the individuals are a *social animal*, i.e. a subject that produces meaningful interactions (tied to her counterparts in a context of norms, values, and meanings), I can state the following: The ethics of responsibility means that the individual can recognize herself in the concept of "common good", that is, a good belonging to each and every individual as members of a community, who can enjoy and pursue it as such. United on the basis of solidarity, that can give meaning to human activity and its development. Sorokin ends *The Ways and Power of Love* (1954) by introducing the dilemma that humanity must face for its survival and development. Humanity can either continue to act according to marauding logics based on egoism (individual and collective) and leading to extinction, or it can embrace logic of universal solidarity that leads mankind to salvation and earthly happiness: each and every individual is given the choice to pursue one path or the other.

3   The Role of Social Sciences and Researchers

Sociology, humanities, and social sciences are an instrument of knowledge of social interconnections because they do not analyze the specific aspects of society as such, but rather their interactions, ties, and reciprocal conditioning. About sociology,

"The central question for sociological theory can then be put as follows: How is it possible that subjective meanings become objective facticities? Or, in terms appropriate to the aforementioned theoretical positions: How is it possible that human activity (*Handeln*) should produce a world of things (*choses*)? In other words, an adequate understanding of the 'reality sui generis' of society requires an inquiry into the manner in which this reality is constructed. This inquiry, we maintain, is the task of the sociology of knowledge" (Berger and Luckmann 1966: 30). Sorokin himself clearly emphasized these features: according to the Russian-American sociologist, "sociology describes only the most common generic forms and stages of development, without, however, pretending to formulate 'laws of development' and 'historical tendencies'" (Sorokin n.d.: Chap. I, pp. 4–5). As Bourdieu had said in his acceptance speech for the CNRS[4] Gold Medal, the task of sociology is "the critical unhinging of the manoeuvring and manipulation of citizens and of consumers that rely on perverse usages of science" (Bourdieu 2013: 12) going beyond the questions posed by common sense or by the media as they are often induced.

In this passage, that leads to abandon a sensuous culture through the affirmation of the ethics of responsibility (or, as Sorokin had prefigured, through creative altruism love or love relationship) the role of researchers (sociologists, psychologists, anthropologists, or any other scholar of humanities and social sciences) is to analyze superorganic phenomena. The end is not their explanation, but their understanding in order to accompany their transformation in favour of a development of a solidaristic humanity.

The work of researchers and the resulting knowledge produced are to be intended, as Bourdieu stated over 20 years ago, in a dual manner: on the one hand, they allow for an "institutional support" (public service) that does not mean meeting all the needs of society, but giving scientific answers to actual problems. Not a "solution", but suggests possible routes for the improvement of the need concerned. On the other hand, they allow the development of a "critical and active citizen" very close to the ideal type of Schütz's "well-informed citizen" (Schütz 1946) which, revised according to the present society (Mangone 2014), appears to be advocating the establishment of a modern citizenship amounting not merely to rights but also to duties. For this newly forged citizenship, the establishment of a socially approved knowledge based on the principle of responsibility (Jonas 1984) becomes a priority, revealed through social reflexivity (Donati 2011a), an aspect of individual reflexivity that is neither subjective nor structural but related to the reality of social relations.

While all work activities produce individual and economic effects, for some of them the implications can also be social and cultural. On the role of sociologists, Ivan Sainsaulieu (2009) refers precisely to these last two characteristics. The French scholar claims that problems related to the role of sociologists cannot be separated from those associated with the engagement and intervention of sociologists in general. In our opinion, reasoning in a logic placing the activities of the sociologist in a

---

[4] The acronym CNRS indicates the Centre National de la Recherche Scientifique, the largest public research organization in France.

relational perspective (Donati 2011a, b; Donati and Archer 2015; Emirbayer 1997), there is no clear distinction between the implications of these activities. There are sociopolitical implications and personal (biographical) ones. And it can be argued, without any doubt, that this *non-differentiation* of implications is real, as social reality consists of objective (objectual) and subjective (symbolic) aspects. Sociology is the search for these real connections, that are both "actions" (intersubjectivity) and "operations" (organizational structure). The boundary between science, profession, and social utility is soon overcome. With regard to the interactions between knowledge and social intervention, if we replace these terms with the concepts of theory, research, and operability, or those of observation-diagnosis-guidance—the ODG-system (see Donati 2011c), we see that they are functionally integrated rather than separated like impermeable environments. We find here again Sorokin's idea of an integrated system of knowledge able to project the activity of sociologists and other social science researchers towards positive social change. And of a sociology—and other social sciences—with a guiding role in these positive changes. After all, researchers "are ordinary human beings who have dedicated their lives to create knowledge" (Valsiner 2017: 25) and who are themselves part of superorganic phenomena.

At this point, we can no longer speak of a contrast between theory and operability. We should rather speak of a continuum of interdependencies that goes from theory to operation, through action-research. Social sciences become the tool to do research. It is indispensable to acquire a knowledge that must "get its hands dirty" to read individual and/or social phenomena, in order to translate theoretical premises into concrete acts. Under this logic, sociology (in particular) and other humanities and social sciences (in general) must play a fundamental role (first) in establishing and (then) in integrating these aspects.

References

Berger, P. L., & Luckmann, T. (1966). *The social construction of reality: A treatise in the sociology of knowledge*. New York: Penguin Books.

Bourdieu, P. (2013). In praise of sociology: Acceptance speech for the gold medal of the CNRS. *Sociology, 47*(1), 7–14.

D'Ambrosio, J. G., Faul, A. C., & Research Fellow. (2014). Love: Through the lens of Pitirim Sorokin. *Analytic Teaching and Philosophical Praxis, 34*(2), 36–46.

Donati, P. (2011a). Modernization and relational reflexivity. *International Review of Sociology, 21*(1), 21–39.

Donati, P. (2011b). *Relational sociology: A new paradigm for the social sciences*. London/New York: Routledge.

Donati, P. (2011c). Cultural change, family transitions and reflexivity in a morphogenetic society. *Memorandum, 21*, 39–55.

Donati, P., & Archer, M. (2015). *The relational subject*. Cambridge: Cambridge University Press.

Emirbayer, M. (1997). Manifesto for a relational sociology. *The American Journal of Sociology, 103*(2), 281–317.

Gehlen, A. (1989). *Man in the age of technology*. New York: Columbia University Press.

Jonas, H. (1984). The imperative of responsibility. In *Search of an ethics for the technological age*. Chicago: Chicago University Press.

Lassman, P., & Speirs, R. (1994). *Weber: Political writings*. Cambridge: Cambridge University Press.

Lassman, P., Velody, I., & Martins, H. (1989). *Max Weber Science as a Vocation*. Boston: Unwin Hyman.

Levin, J., & Kaplan, B. H. (2010). The Sorokin multidimensional inventory of love experience (SMILE): Development, validation, and religious determinants. *Review of Religious Research, 51*(4), 380–401.

Mangone, E. (2014). La conoscenza come forma di libertà responsabile: L'attualità del 'cittadino ben informato' di Alfred Schütz. *Studi di Sociologia, 1*, 53–69.

Melucci, A. (1994). *Passaggio d'epoca*. Milan: Feltrinelli.

Sainsaulieu, I. (2009). Il coinvolgimento del sociologo nel suo oggetto: Il caso del lavoro sociale, sanitario e di cura. *Salute e Società, 8*(suppl 3), 133–148.

Schütz, A. (1946). The well-informed citizen. An essay on the social distribution of knowledge. *Social Research, 14*(4), 463–478.

Sorokin, P.A. (n.d.). *The nature of sociology and its relation to other sciences*. University of Saskatchewan: University Archives & Special Collections, P.A. Sorokin fonds, MG449, I, A, 3.

Sorokin, P. A. (1928). *Contemporary sociological theories*. New York: Harper.

Sorokin, P. A. (1948). *The reconstruction of humanity*. Boston: The Bacon.

Sorokin, P. A. (1950). *Altruistic love: A study of American good neighbors and christian saints*. Boston: Beacon.

Sorokin, P. A. (1955). Les travaux du Centre de recherches de Harvard sur l'altruisme créateur. Cahier internataux de. *Sociologie, 19*, 92–103.

Sorokin, P. A. (1954). *The ways and power of love: Types, factors and techniques of moral transformation*. Boston: Beacon.

Sorokin, P. A. (1958). Integralism is my philosophy. In W. Burnett (Ed.), *This is my philosophy. Twenty of the world's outstanding thinkers reveal the deepest meaning they have found in life* (pp. 180–189). London: George Allen & Unwin.

Valsiner, J. (2017). *Mente e culture: La psicologia come scienza dell'uomo*. Rome: Carocci editore.

Chapter 8
Epilogue: Towards Integral Social Sciences

The verb "to conclude" comes from the Latin *cum* and *claudere* and means "enclosing", "closing in", but these meanings do not fit well with all the reflections and musings here presented. There is no end to any discourse on Sorokin and the integral method. I rather wish to emphasize here a condition of dynamism, of "openness" among the disciplines falling under the label of humanities and social sciences. No theorization is concluded or coming to a close. Starting from the dialogue between different disciplines, we can imagine a propulsive push towards the integration of knowledge systems from the less complex to the more complex ones, and vice versa.

Summing up Sorokin's thoughts on the integral approach is a very difficult task due to his scientific production of many hundreds of pages. In it, although not directly, the issue of knowledge and of the means to reach it is always discussed. The purpose of knowledge is to provide a more complete and valid picture of social reality for the understanding, interpretation, and prediction of sociocultural phenomena. For this reason, the graphic representation (Fig. 8.1) I propose here is a result from my personal interpretation.

Assuming the three indivisible components (society, culture, and personality) of sociocultural phenomena—something that Sorokin clarifies in *Society, Culture, and Personality* (1962)—and the fact that every sociocultural phenomenon is characterized by a spatial and temporal dimension—another aspect clarified by Sorokin in *Sociocultural Causality, Space, Time* (1964)—we will describe and interpret the graphical representation of how an integrated system of knowledge is produced. And how the latter, if realized, can promote the passage from the order of explaining (*erklären*) to the order of uderstanding (*verstehen*). The search for the reason of sociocultural phenomena should no longer refer to a *cause*, but to a *meaning* that can be the key to interpret the dynamics of individual-society interactions.

There are two necessary premises: (1) Time and space are both constitutive elements of social interaction processes, but they are also two central categories of and for sociological analysis, and for social sciences in general. The daily experiences of individuals can be perceived—and therefore studied—in their continuous

© The Author(s) 2018
E. Mangone, *Social and Cultural Dynamics*, SpringerBriefs in Psychology,
https://doi.org/10.1007/978-3-319-68309-6_8

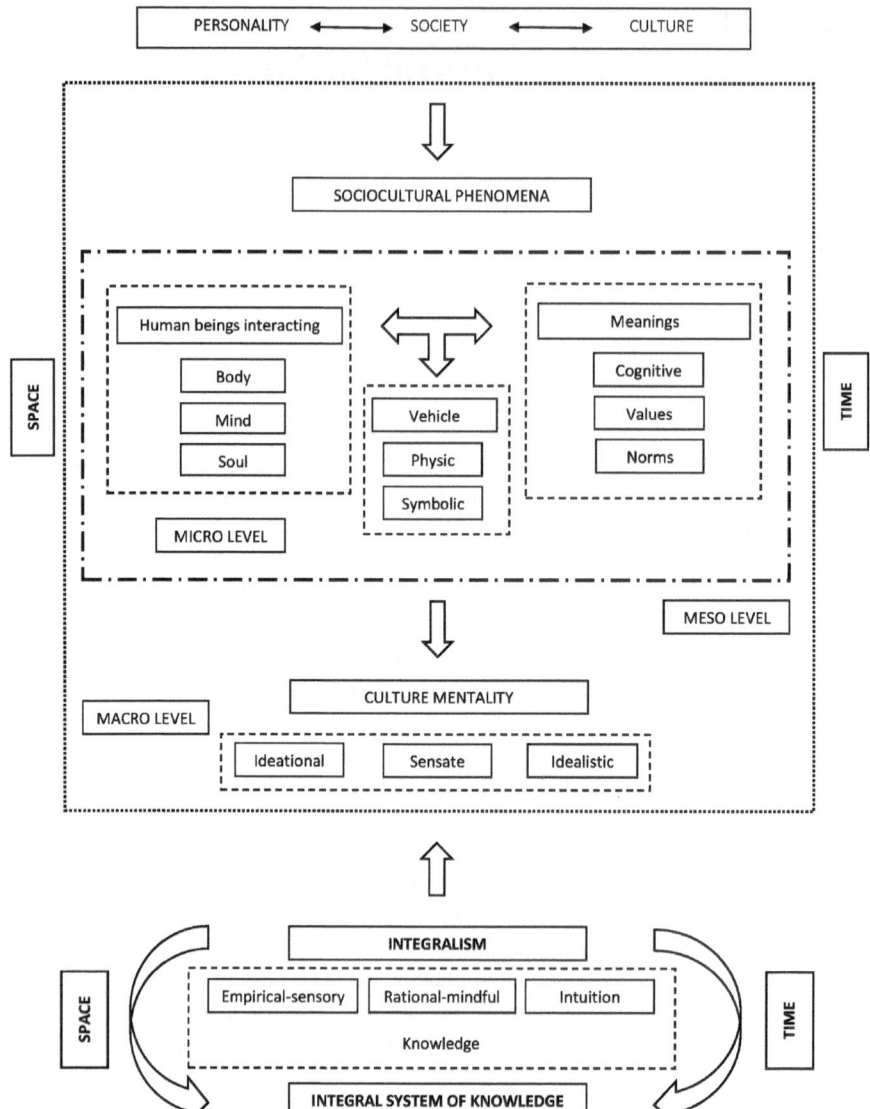

Fig. 8.1 Graphic representation of sociocultural phenomena and of the integral method

unfolding (observation), as they flow into the unity of the individual experience and of the situation (the scripts presented above); or they may become the subject of a subsequent reflection when we reflect on them after they have been experienced.

In the first case, personal history coincides with experience—and this also applies to the researcher—and cannot be separated from it; in the second, instead, reflecting on past actions means that they are considered disconnected from personal experience. Thus, time is no longer unitary and being aware of this means that individuals are oriented in their actions/interactions by the temporal and spatial dimensions (social and historical context), and the same happens to the activities of the researchers; (2) The study of sociocultural phenomena must consider the three different levels of analysis of sociology and of social sciences in general: (a) the *macro-social* level (concerning social systems and their forms of organization) that in Fig. 8.1 is represented by the study of the cultural mentalities identified by Sorokin (Ideational, Sensate, and Idealistic); (b) the *micro-social* level (relating to the individual–society relationship and social actions), represented in Fig. 8.1 by human beings interacting (body, mind, and soul); (c) the *meso-social* level (concerning the relationship between the social system and the world of life, where the latter is understood as the set of meanings and representations of culture), in Fig. 8.1 it is represented by the meanings (cognitive, values, and norms) and vehicles (physics and symbolics) that allow human interaction.

Keeping together the three levels of analysis described above implies an intellectual action that goes beyond the "disciplinary" points of view and the methods of investigation (qualitative and quantitative). The study of sociocultural phenomena and the methodologies adopted to this end must be oriented towards the integration of the subjective and objective dimensions. The element holding it all together is the interpretation and construction of reality through the relationships between individuals, and between them, society, and culture. Since individuals are interacting agents (in the world of everyday life and in institutions) all these aspects should be read as a *correlation of interpretations* and not just as the answer to a triggering cause.

For the study of sociocultural phenomena, it is therefore necessary to consider an integrated mix of factors, disciplines, and investigation methodologies. Sociological knowledge and that of other social sciences must be integrated into a single integrated system of knowledge (integral social sciences) that must focus its attention on all aspects of the transformation of society (holistically: aspects of personality, society, and culture) without neglecting reflections on the activities of the researcher herself. The solution proposed by Sorokin is Integralism, a system of knowledge that integrates the three forms of knowledge (empirical-sensory, rational-mindful, and supersensory-superrational) acquired via three means (sense, reason, and intuition). These forms of knowledge have no intrinsic hierarchy; rather, they are all indispensable—for this reason, in Fig. 8.1 they are represented on the same line.

An integrated system of knowledge thus configured must be associated with action, so as to allow us to seek not the solution, but the possible paths to undertake to improve the social problems considered. The work of social scientists is therefore one that has a "weight", as they are themselves actors in society. As such, they bear values and meanings, as well as subjective and social rights that must be taken into account during the research activity for the purpose of creating new knowledge on sociocultural phenomena: "Scientists are not machines, nor soldiers in the massive army of science ready to attack yet another fortress of ill-being in the world: cancer, malaria, and any other threatening illness or social condition" (Valsiner 2017: 25). Thus, this system of knowledge is produced by the relation/interaction between the researcher and the object of her inquiry, between self and other, without dependencies or hierarchical levels of sorts.

Because of the rapid succession of changes in society, it is also necessary, in light of what has been said here, to consider the role of social sciences in reading social transformations. Today, while a new way of thinking that even involves the organizational structures of the world's major institutions is developing, social sciences seem to have difficulties in reading these transformations. These difficulties are probably due to positions of excessive self-referentialism in the disciplines. But it is precisely in this continuously changing context that social sciences can play a paramount role as sciences able to understand society. This, however, requires researchers to redefine paradigms, methodologies, and methods, so that knowledge is configured as a network experience. This form of knowledge is the result of comparisons and conflicts occurring in a specific space and time in an integrated system of knowledge—something that Sorokin had already suggested many years ago, but that, of course, fell on deaf ears.

Through a systematic and methodologically founded observation—seen as the main activity to overcome the Comtian "social physics"—we can lay the foundations for building an integrated system of knowledge derived from an integral social science. It is therefore necessary to redefine the paradigms of sociology and other social sciences in a direction that keeps together the different dimensions (macro, meso, and micro). On these three aspects, we must keep open the door to free and autonomous scientific reflection. The latter, however, necessarily needs to make a leap towards operability. It does not have to provide the answers, but the guidelines and *tools* that serve as a guide for policies (welfare, education, economic ones, etc.) to be implemented in order to concretely realize a society that is more open and more "fit" for individuals.

It is therefore desirable that an integral theory of knowledge develops and becomes reflective knowledge able to promote positive interactions in the living environments of individuals and between individuals. The autonomy of the individual disciplines of social sciences (sociology, psychology, anthropology, etc.) must not be neglected, but we must abandon the excess of self-referentiality that limits all available knowledge within the reference cadres and paradigms of individual disciplines.

References

Sorokin, P. A. (1962). *Society, culture, and personality: Their structure and dynamics, a system of general sociology*. New York: Cooper Square.
Sorokin, P. A. (1964). *Sociocultural causality, space, time: A study of referential principles of sociology and social science*. New York: Russell & Russell.
Valsiner, J. (2017). *From methodology to methods in human psychology*. Geneve: Springer.

Index

A
Accommodation process, 56
Altruism
    in Sorokin, 73–76
Altruistic love, 73–76
Aristotle's study, 9
Ascetic Idealionalism form, 57
Assimilation process, 56

B
Bioconscious, 65
Boudon's Weberian paradigm, 46

C
Causal/functional integration, 32
Choice
    forgotten theories, 1
    personal and family events, 2
    reasons, 1
    silent scream, 2
    student, 1
    traditional approaches, 2
Collective intelligence, 31
Conduct. *See also* Human conduct
    individual's, 64
    social mentality, 63
Consensual universe, 31
Contemporary Sociological Theories (book), 74
Contemporary Sociological Theory, 34
Culture
    behavioural, 54
    habitus, 59
    human culture, 53
    ideological, 54

    material, 54
    mentality of, 56–58
    objective elements, 53
    subjective elements, 53
Cynical Sensate Mentality, 58

D
Durkheim's sociological analysis, 12
Dyad knowledge/reality, 29–32

E
Ethics
    responsibility, 76–78

F
*Fads and Foibles in Modern Sociology and
        Related Sciences*, 38
Frankfurt School, 14, 22

G
Giddens' theory, 15
Gnoseological problem, 30
Goffman's contribution, 65
Goldthorpe's criticism, 17
Griswold's cultural diamond, 60

H
Habermas's theory, 15
Human conduct
    trust and risk, 68–71
    uncertainty, 67, 68

© The Author(s) 2018
E. Mangone, *Social and Cultural Dynamics*, SpringerBriefs in Psychology,
https://doi.org/10.1007/978-3-319-68309-6

I
Idealistic culture mentality, 58
Individual
    biographies of, 45
    constant oscillation drives, 66
    culture, 54
    development, 64
    dialectical process, 44
    experience knowledge, 55
    identity, 66
    individual–body relationship, 46
    individual's actions, 58
    inner change, 58
    inseparability of, 64
    Lebenswelt, 61, 64
    personality, 66
    social and cultural components, 55
    social definition, 65
    social lives of, 45
    sociocultural interactions, 42
    transformation of, 66
Indivisible sociocultural trinity, 49
Integral theory of knowledge
    Dyad knowledge/reality, 29–32
    integralism, 35–39
    Sorokin's sociology of knowledge, 32–34
    systems of truth, 37–39
Integralism, 35–39, 48, 85
Internal/logico-meaningful unity, 32

J
Journey
    intellectual, 5
    Sorokin, 5–7

L
Laminal model, internalization/
            externalization, 48
Linguistic sign, 30

M
Macro-social level, 85
Meanings
    cultural signifies concentrations, 42
    objective meanings, 70
    provinces of, 64
    Schütz's theory, 67
    social, 42
    sociocultural phenomena, 44
    symbolic, 43
Mental consciousness level, 38

Mentality, 34
Meso-social level, 85
Micro-social level, 85
Moscovici's monumental theoretical work, 22

O
Observation-diagnosis-guidance (ODG)-
            system, 80

P
Paradox
    contemporary society, 49–51
Parsons' functionalism, 11
Personality
    individual (identity), 66
    supra-conscious and conscious, 66
    Weltanschauung, 63–67
Pseudo-ideational culture mentality, 58
Psychoanalytic therapeutic practices, 23
Psychology
    American debate, 18–21
    cultural, 24
    and economics, 17
    sociology, 13, 21–23
    subjective, 16

R
Research
    budding sociological, 11
    empirical, 17
    interdisciplinary, 23
    subject of, 11
Researchers, 78–80
    activities of, 85
Responsibility
    ethic of, 76–78

S
Schütz's theory, 67
Social problems, 59–61
Social sciences
    American, 4
    functions of, 2
    humanities, 83
    individual disciplines of, 86
    integral, 85, 86
    quantophreny of, 2
    researchers, 78–80
    role of, 1
    sociology, 85

Social structure, 43
Society
    contemporary, 49–51
    social relations, 45–49
    social universe and interactions, 41–45
Socioconscious, 65
Sociocultural phenomena, 11
    graphic representation, 84
Sociological imagination, 38
Sociology
    American debate, 18–21
    crisis of, 18
    critical, 7
    development, 9–15
    evolutionary stages, 10–13
    knowledge, 13–15
    object and boundaries, 15–18
    practicing, 2
    and psychology, 21–23
    role of, 1
    Russian-American sociologist, 1
    sociological epistemological framework,
        18
    sociologistic school, 3
    transdisciplinarity, 23–25
Sorokin's American period, 4, 5
Sorokin's concept, 69
Sorokin's indivisible sociocultural trinity, 47
Sorokin's integralism, 29
Sorokin's integrated system of knowledge, 48
Sorokin's philosophy, 37
Sorokin's reflection, 59
Sorokin's Russian period, 3, 4
Sorokin's sociology of knowledge, 32–34
Sorokin's theoretical framework, 45
Sorokin's theory, 63
Sorokin's works, 5, 7
Spatial/mechanical adjacency, 32
Superorganic phenomena, 6, 10
Superorganic phenomenon, 55, 59–61
Supersensory-superrational intuition, 39

T
Talcott Parsons' works, 22
The Crisis of Our Age, 41, 49
Thomas Kuhn's work, 13
Transdisciplinarity, 23–25

U
Uncertainty, 67, 68

W
Weltanschauung, 57, 63, 67, 77
Wissenssoziologie, 31, 32